W9-ATW-892

Pelican Books
Geology and Scenery in England and Wales

Sir Arthur Trueman was educated at High Pavement and University College. Nottingham. He was Professor of Geology successively at Swansea. Bristol and Glasgow. but resigned his chair in 1946 to become Deputy Chairman of the University Grants Committee. of which he became Chairman in 1949.

Besides his interest in the development of scenery. his researches ranged from the more philosophical aspects of the study of fossils to the investigation of the British coalfields; in this latter sphere he established a basis for the identification of coal seams which has been applied to many British and European coalfields. He published *An Introduction to Geology. This Strange World* and other books. and numerous papers. He was elected a Fellow of the Royal Society in 1942. was President of the Geological Society (1945–7). Chairman of the Geological Survey Board. and a member of the Advisory Council on Scientific Policy. He was much interested in the colonies and visited West Africa with the Elliot Commission in 1944. He was created K.B.E. in 1951. and died in 1956.

Dr John B. Whittow is a Senior Lecturer in Geography at Reading University. His interests are in the geomorphology of highland Britain and landscape evaluation. He has held posts in Africa. North America and Australia.

Dr John R. Hardy is also a Lecturer in Geography at Reading University. His main research interests are coasts and coastal processes. A graduate of Cambridge. he has held posts in Canada and the U.K.

Cover photograph. *The Dorset coast west of Durdle Door.*

Frontispiece. *The scarps of the Welsh Borderlands, looking north-eastwards from Craven Arms, Shropshire. The scarp on the left is Wenlock Edge, an outcrop of Wenlock Limestone, that on the right is Callow Hill, formed by the younger Aymestry Limestone.*

GEOLOGY AND SCENERY IN ENGLAND AND WALES

A. E. TRUEMAN

New edition completely revised by
J. B. Whittow and J. R. Hardy

PENGUIN BOOKS

Penguin Books Ltd, Harmondsworth,
Middlesex, England
Penguin Books, 625 Madison Avenue,
New York, New York 10022, U.S.A.
Penguin Books Australia Ltd, Ringwood,
Victoria, Australia
Penguin Books Canada Ltd, 41 Steelcase Road West,
Markham, Ontario, Canada
Penguin Books (N.Z.) Ltd, 182–190 Wairau Road,
Auckland 10, New Zealand

First published as *The Scenery of England and Wales* by Gollancz
1938
Published in Pelican Books 1949
This revised edition first published in Pelican Books 1971
Reprinted 1974, 1977
Copyright © the Estate of A. E. Trueman, 1948
Revisions copyright © J. B. Whittow and J. R. Hardy, 1971

Made and printed in Great Britain by
Butler & Tanner Ltd, Frome and London
Set in Lumitype Janson

To E. R. T.
Who made me write this book

Contents

Acknowledgements

The publishers wish to thank the following for permission to use their photographs: Aerofilms Ltd for the Frontispiece and Plates 2, 5, 6, 12, 14, 15, 21, 22, 25, 28, 29, 30 and 31: copyright reserved; Blakes of Hanley, Stoke-on-Trent, for Plates 16 and 17; the British Tourist Authority for Plates 4, 8, 11, 13 and 32; Mr Wilfred Elms for Plate 24; Dr Kathleen Simpkins for Plate 23. Plates 3, 7, 9, 10, 18, 19, 20, 33, 34 and 35, Crown Copyright Geological Survey photographs, are reproduced by permission of the Controller of H.M. Stationery Office. Plates 26 and 27 by J. K. St Joseph, Cambridge University Collection: copyright reserved.

The map on the back cover and Figures 9, 25, 28, 29, 30, 34, 40, 41 and 43 are based on Institute of Geological Sciences maps, and are reproduced by permission of the Director.

Some diagrams were drawn by Mr H. Stewart-Killick; the others were drawn and adapted by Mrs Janet Dore and Mrs Mary Petts. The map on the back cover was drawn by Penguin Education Art Department.

List of Plates

Preface to the First Edition

In this book an attempt is made to describe the scenery of most of
England and Wales, and to show how some of the features have
come into being. It has been prepared because I believe that there
are many people, motorists and walkers as well as students, who
desire this kind of information, and because singularly few works
on the subject are available. The recent spate of topographical
works of one type or another, many of them of surpassing excel-
lence, is evidence of the great interest taken in the English countryside
by increasing numbers of people. And yet the scenic background of
the village and church has received little attention.

The present volume is written partly though not entirely from
the geological standpoint. I have tried to make my account intel-
ligible to all who may be interested. There is no preliminary discussion of principles, and few technical terms are introduced. These
are explained simply in the text as they are used, and rather more
precisely in the glossary. But in any attempt at popularization
certain risks must arise, and I trust that my efforts at simplification
have not led me to be too dogmatic or too general in my statements.

The work was started light-heartedly during a summer vacation;
much of the earlier part was written in a quiet bay on the Cardigan
coast. As the writing has proceeded I have become more and more
conscious that what I have written falls short of what I had planned
and I can only hope that the result is not wholly unworthy of the
theme.

I hope that the broad ideas on geological history which will be
gained from the reading of this book will encourage a wider
appreciation of the science. For geology is pre-eminently the
layman's science. In it more than in any other science there is

opportunity for a beginner to make original observations, to weigh up evidence, to coordinate his facts, and in general to acquire a truly scientific outlook, whereas a layman can do no more in many sciences than accept ready-made conclusions, often explained by clever but dangerous analogies, without any prospect of understanding the steps by which they have been reached.

Each chapter deals with one area or type of country, and they are so arranged that the more complex areas are described last. The more serious student will find it advantageous to read the chapters in order, but I think that the general reader will be more interested if he first turns to the areas with which he is familiar. It should be made clear, however, that the accounts of the various districts are not always complete in themselves, as certain features are best dealt with when other regions are being considered.

December 1937   A. E. TRUEMAN

Preface to the Revised Edition

Sir Arthur Trueman's excellent book is still in great demand some thirty years after its original publication, which must mark it as a near classic in its field. However, with advances in geology and allied sciences, and with improved printing techniques, it was thought that a revised edition was necessary, if the book was not to fall too far behind the times.

So the text has been revised to take account of thirty years of new work, and photographs have been introduced to replace Sir Arthur's original line drawings. Almost inevitably, revision has made the text longer. Conscious that the original edition was read mainly by interested laymen and also by students in the sixth forms of schools and those beginning at university, the revision is aimed at the same types of reader. Some new material and concepts have been introduced, and it is hoped that these will readily be grasped by the amateur; the precepts of the first edition, as described in its preface, have constantly been kept in mind.

Where possible Sir Arthur's characteristic style and descriptions have been retained. Some chapters have been left with comparatively minor alterations, while others have been almost entirely re-written to take into account the results of recent research in particular regions. A new chapter has been added on the lowlands of north-west England; Devon and Cornwall have been incorporated into a single chapter, and minor alterations have been made in the order of chapters. One of us (J. R. H.) has been primarily responsible for the south and east (Chapters 2–9 and 21), while the other (J. B. W.) has dealt with the north and west (Chapters 10–20), a division which corresponds almost to that of 'lowland' and 'highland' Britain.

J. B. W.

J. R. H.

1. The Foundations of Scenery

In England and Wales we are singularly placed to appreciate the relationship of scenery and structure, for few other parts of the earth's surface show in a similar small area so great a diversity of rock types and of landscape features: 'Britain is a world by itself'; its mountains are not high, nor its rivers long, but within a few hundred miles of travel from east to west one may see more varieties of scenery than are to be found in many bigger countries.

There are hill-forms of every type; smooth-contoured chalk downs in a great belt encircling the London basin, the long line of the Cotswold edge, even-topped plateaux in Wales and Devon, bleak granite moorlands and craggy volcanic cliffs. All these types reflect their underlying structure and their origin is apparent in their shape.

In many of these different regions the native stone has given character and fitness to the buildings; the wide variety of stones, together with the frequent use of wood and brick and thatch, especially in the stoneless areas, has added a charm to English villages and towns, and has made the dwellings, particularly the smaller cottages (for stones for churches and large houses were carried for many miles even in early times), ideally suited to their surroundings. With the improvement of transport during the nineteenth century, building materials were more frequently carried over greater distances, and soon after the opening of railways houses began to lose that harmony with the countryside so characteristic of the older cottages. The towns which thus sprang up at some railway centres are examples of the way in which individuality was lost, and lately the housing estate has spread a new uniformity even more widely over the country.

It may be suggested that a traveller can appreciate these aspects of landscape without knowing anything of their anatomy, and that a geologist who must spend much of his time searching the ground is no fit guide for those who seek to know the countryside. It must be urged, however, that many never know an area until something of its meaning becomes clear to them, just as some may never feel comfortable in a strange town until they have seen a map of its streets. For a country is not just a jumble of hills and valleys; the features have a plan, a system underlying their distribution, and once this is understood the region is seen more clearly and its variety more readily appreciated. The geologist acquires an eye for country and an understanding of nature not excelled by that of the artist or the poet.

The areas described in the following chapters are chosen to illustrate the diversity in form and colour of different parts of England and Wales. The Cotswolds and comparable areas are dealt with first because they present a simplicity of structure which is not found in any other large region. The chalklands have some features in common with the Cotswolds but their distribution introduces new problems of structure, which lead to a discussion of the London basin and the Weald, each bounded by chalk regions, and of the south coast where the Hampshire basin may be likened to the London basin partly sunk beneath the sea. At this point, after a discussion of the landscapes of East Anglia and the Fenlands, the Midlands are described in some detail. The latter is in many senses a transition area; a zone in which the old, hard rocks, older than the Coal Measures, begin to protrude through the newer sedimentary rocks which blanket them in southern and eastern England. The Midlands also mark the zone where one passes from 'lowland' into 'highland' Britain, and finally they illustrate the scenic change as one travels north into an England where the Industrial Revolution has left its violent mark on the landscape. Although this is not to say that south-eastern England was unaffected by such change, it was on the coalfields that the greatest impact was felt. The mineral wealth of the rocks was the most important single factor which determined the location of many of Britain's industrial towns.

The Pennines, the Lake District, Wales and the West Country are taken in the later part of the book since each area introduces complexities of structure, missing from the regions so far described, not least of which are the concepts of igneous rock formation.

In the description of these areas some reference is made to their geological history. Most of them are made up of sheets of sedimentary rock laid down on the floor of some ancient sea, but in some there are also sheets of lava and other rocks of igneous character. In general the scenery is controlled by the nature and form of these various beds, many of which have been bent, crumpled and broken at different times owing to the great movements in the earth's crust, the edges of the sheets thus reaching the surface at various angles.

These sediments were laid down in the seas which at different times have covered the area which is now Britain. Of the antiquity of the rocks there is no doubt, but it is not sufficient merely to think of all these rocks as having been formed millions of years ago. They represent a sequence of events, and their histories can be related to a time-scale. But there is no need to worry about putting our time-scale into years; it is sufficient to realize that an area composed of very ancient rocks may have undergone many changes since the rocks were first formed, and though the surface features are built up of the old bed-rocks many of them may be of comparatively recent origin. Most of the river valleys of England and Wales have been formed within periods which from a geological point of view are comparatively recent — some tens of millions of years at most. Almost at the end of that time, and finishing not more than a few thousand years ago, the great Ice Age led to important modifications in the surface of many parts of Britain.

In the description of most areas little reference is made to this question of a time-scale, but in order that the more scientifically inclined reader may get a more precise knowledge of the relations of the various episodes which are touched upon here and there, a brief account of the chronology is added as a final chapter (Chapter 22), and frequent reference to the diagram on page 356 may help to indicate more clearly the sequence of events.

2. The Cotswold Stone Belt

The Cotswold Hills, stretching north-eastwards from near Bath and merging, sixty miles away, into the uplands of Northamptonshire, form an area of essentially consistent character. In this area the land rises in few places above 1,000 feet, but the general altitude is between 400 and 800 feet. Rising gradually from the low vales of Oxford, the greatest heights are near the western boundary, where the hills fall away rapidly to the Midland plain and the Vale of Severn. This steep scarp face is indeed the most impressive feature of the Cotswolds and forms a boundary of unmistakable distinction.

Travelling westwards by coach about 150 years ago, Sydney Smith was impressed by this contrast, and his forceful description emphasizes at least one essential fact:

The sudden variation from the hill country of Gloucestershire to the Vale of Severn, as observed from Birdlip or Frocester Hill, is strikingly sublime. You travel for twenty or five-and-twenty miles over one of the most unfortunate desolate counties under heaven, divided by stone walls, and abandoned to the screaming kites and larcenous crows: after travelling really twenty and to appearance ninety miles over this region of stone and sorrow, life begins to be a burden, and you wish to perish. At the very moment when you are taking this melancholy view of human affairs and hating the postilion and blaming the horses, there bursts upon your view, with all its towers, forests and streams, the deep and shaded Vale of Severn.

We need not say that Smith was unkind to the Cotswolds in this account, but he doubtless shared with many of his time a dislike for upland country. It is not necessary, however, to defend the Cotswolds against his charges. This is an area of stone though not of sorrow; indeed, in illustration of the change of attitude to such uplands, we may note that a more recent traveller (J. B. Priestley) has felt that the Cotswold walls know 'the trick of keeping the lost sunlight of centuries glimmering upon them'.

The Cotswold Hills as well as their villages are built of stone, mostly of the light-coloured limestone which weathers to rich and varied tints of brown and yellow. Often these rocks are near the surface, and the light-brown soils in the fields are thin and made up of a rubble of pale limestone fragments. In many areas the fields are divided by dry-stone walls, dug long since from shallow quarries which often extend along the roadside. The buildings likewise are almost universally of stone, and the whole area has a fitting colour and pattern that cannot but be a delight to modern eyes. A single building of brick here becomes an eyesore.

It is clear, then, that the essential characters of this area are determined by the rocks. These limestones are to be seen in innumerable small quarries, and in the stone walls themselves. In many parts they are full of shells of various types, together with corals in some cases. They certainly represent deposits laid down on an ancient sea-floor, and it is apparent that a wide area of such a sea-floor has been elevated into these hills, for rocks of similar type (and in fact of equivalent age) cover the hill-tops over the whole region. Where quarries make the arrangement of the rocks visible, it is seen that they occur in beds lying more or less horizontally: actually the beds are tilted very slightly, and if they are examined over a sufficiently wide area they will be found to tilt gently towards the south-east, although locally they may be inclined westwards: when they were upraised from beneath the sea they were lifted higher on the north-west than on the south-east, the minor rolls resulting from irregularities in the original uplift or from slight changes at subsequent times.

Many of the limestones possess a structure which is known as oolitic, and these rocks are frequently spoken of as the oolites. Such oolitic limestones consist mainly of minute, rounded spheres of calcium carbonate, massed together and resembling the roe of a fish, from which they derive their name. Many such limestones show an oblique lamination, indicating that the rounded grains have been arranged by strong currents in shallow water, and it may be taken as generally true that the limestones of this area are of shallow-water origin. But not all are oolites: it is obvious that some are composed of shells or shell fragments. The nature of the stone, as well as the thickness of the beds, has controlled its use for building. The shelly limestones break irregularly and in general can only be used for rough walling, while some of the thicker oolite beds are much more readily dressed into blocks of any required shape. The quarrymen thus recognize 'freestones' (which can be so dressed) and 'ragstones'. Many of the freestones can be cut easily when freshly quarried, but harden on exposure. The thinner beds of limestone are used largely for roofing purposes, and are locally known as 'slates', for example the Collyweston Slates, though of course they differ greatly from the true slates of Wales. It is this association of such varied types of rock within the Jurassic series (Jurassic being the name given by geologists to a long period of past time, see Chapter 22) that makes possible the building of cottages, mansions and churches from stone of local origin (Plate 2).

The most striking feature of the Cotswolds is the general form: the steep scarp slope on the north-west, and the long, gentle dip slope down to the clay vales on the south-east. The scarp often covers less than a mile, in which the ground may rise six or eight hundred feet, while the dip slope to the south-east may be as much as twenty miles in width. The great asymmetry of the hills is due to their geological structure, being controlled by the gentle dip or tilt of the limestone beds which, as has already been stated, is to the south-east. The dip of the limestone is, however, generally greater

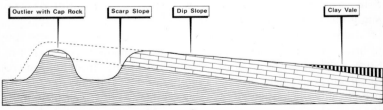

Fig. 1. *A receding escarpment with an outlier.*

than the slope of the ground surface and thus, travelling down the dip slope, one crosses successively younger rock outcrops (Fig. 1). These younger rocks, which lie above the oolites, are here mainly clays, forming the vale around Oxford. At the scarp, how-ever, it is the hard limestones which form the summit of the scarp, due to their resistance, while softer rocks beneath are undercut to form the steep scarp slope. Minor hard bands can repeat the major scarp structure on a smaller scale on the main dip slope.

Many other hill ridges in central and southern England are of this character: a hard bed of sandstone or limestone between softer beds producing such an asymmetrical hill, or scarp. Indeed, the whole area of the Cotswolds and their extensions north and south from Yorkshire to Dorset, and the land south-east as far as the chalk scarp, is frequently called the English scarplands, because this scarp and vale structure is repeated several times in different rocks.

William Smith, one of the founders of geological science, likened this type of structure to slices of bread and butter arranged on a plate, a helpful comparison which must not be carried too far, for there is nothing different or permanent about the crust of the slices. The cyclist crossing the country used to be well aware of these different slopes: going from west to east he had short but steep slopes to climb, with long easy runs down the dip slopes, while on his return he had long dreary climbs, with steep descents that were often too sharp to be ridden in safety. True, improved roads have to some extent eased out many of these gradients, but the asymmetry of the hills is still marked enough to impress itself on the motorist, the stiff gradients of the Cotswold edge, notably at Birdlip and Broadway, being well known (Plate 1).

The attack of the weather along the steep slope wears away the softer clays more rapidly than the limestones; moreover, the rain which falls on the limestone mostly sinks underground, scarcely affecting the limestone surface, while that which falls on the clay and that which subsequently rises in springs along the junction of the two flows over the clay and carries some clay continuously downhill. The lower parts of this steep slope therefore tend to be always concave, the clay being to some extent hollowed from underneath the limestone. Those who drive or cycle up the face of

Plate 1. *The Cotswold scarp at Birdlip Hill, Gloucestershire. The view looks north-west down the scarp slope and across the Severn valley. An outlier, Churchdown Hill, can be seen in the middle distance; this is capped by Upper Lias rocks, while the vale is floored by Lower Lias clays. The scarp slope itself is made up of Middle and Upper Lias, and Inferior Oolite rocks. The distant skyline from this viewpoint is formed by the Malvern Hills, the Woolhope dome and the Forest of Dean.*

such a hill will be familiar with this type of change of slope, at any rate on old roads, although many newer roads are more carefully graded.

The continuation of these processes leads to the limestone being so far undermined that in time portions may slip down the slope, to be broken up gradually, while weathering processes attack a new surface of clay and again produce the concave form of hill. In short, any position of the edge is only temporary, and, although very slowly, the effects of denudation have caused it to recede to positions farther south-east than those it once occupied. It will be apparent, therefore, that when speaking of the Cotswold limestones as representing a former sea-floor which has been upraised, it must not be thought that the present border of the Cotswolds marks the limit either of the sea-floor or of the uplift. The possible extent of the Cotswolds at an earlier stage need not be discussed, but clearly if the Cotswold uplands extended far to the west, there was at that time no Vale of Gloucester, a point to which reference may suitably be made later.

The fact that the Cotswolds formerly extended farther to the west is shown by the existence of a number of detached outliers of the main ridge (Fig. 1). One of the most notable of these is Dundry Hill, an elongated hill about four miles south of Bristol, which is separated by six miles from the present limestone edge to the north-west of Bath. This outlier of Dundry shows that the whole of the area between Bristol and Bath must once have been covered by the oolitic limestones: moreover, Dundry shows the characteristic scenery and building stones of a Cotswold area though it is cut off by the other rock types which surround it. From the Dundry quarries was obtained stone for the beautiful church of St Mary Redcliffe, in Bristol. Scenically and agriculturally as well as geologically, Dundry is an outlier of the Cotswolds.

Small outliers, though even more striking, rise sharply out of the plain near Gloucester to form Robin's Wood Hill and Churchdown, while farther north again the large mass of Bredon Hill, between Evesham and Tewkesbury, is still more conspicuous. Such an outlier is shown in section in Figure 1.

The view from the Cotswold edge across the valley to the west is always impressive. From Frocester Hill near Stroud and from Leckhampton Hill near Cheltenham the view embraces the wide clay plain, and its outlying oolitic hills, with the Severn widening out into its estuary, and beyond, the irregular crest of Malvern and the round mass of May Hill with its crown of trees, marking the real boundary of the English plain.

The face of the scarp is often wooded: beech is the commonest tree on these dry soils, but pines have also been planted in some areas. Cranham Woods north of Painswick, traversed by the road from Cheltenham to Stroud, afford many beautiful glimpses over the plain. But though the scarp is steep, it varies greatly in pattern. From Old Sodbury for some eight miles it extends due north in almost a straight line, but thence to Dursley and Stroud the upland is deeply curved and its edge is fretted into long, winding valleys of striking beauty. Here too the edge is complicated by the presence of a hard band of ferruginous limestone (Marlstone) which occurs among the Lias clays underlying the oolites. Near Gloucester and Cheltenham the scarp is rather simple in pattern, but as Worcestershire is approached it is cut by the wide embayment of the Vale of Moreton. There is little marsh hereabouts but Moreton-in-Marsh aptly suggests the great southward extension of the clays which floor the Vale of Gloucester and the Vale of Evesham. The Vale of Moreton separates the Cotswolds from the Northamptonshire uplands to the north-east.

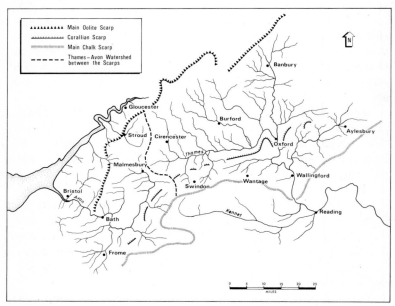

Fig. 2. *The drainage of the Cotswolds.*

The drainage of this Cotswold tract is shared by the Thames, the Bristol Avon and their tributaries. The Thames system drains the larger north-eastern part, while the Avon drains the southern portion (Fig. 2). The bulk of the smaller streams flow from north-west to south-east or west to east, and many of these follow the general direction of the dip of the beds, and they are referred to as dip or consequent streams, since their direction is a consequence of the inclination of the beds on which they flow. Subsequently to this, other streams carve out a course for themselves transversely along the line of a weaker bed; these are known as strike or subsequent streams (Fig. 3). Examples of these can be found, notably the upper Thames itself, draining the Oxford clay vale (Fig. 2). The lower Avon may well have been a large dip stream in earlier times when the scarp was much farther to the west, and it may once have contributed to the headwaters of the Kennet (Fig. 2). Its direction of flow was probably reversed late in geological time, with the downfaulting of the Bristol Channel (see p. 325). The smaller dip streams occupy wide and generally shallow valleys on the upland surface. Indeed they introduce the chief diversity into the uplands, for along many of them extend belts of woodland and in them are situated many of the villages; the availability of water supply was a prime factor in determining the location of many of the early settlements. Many valleys are dry, but nevertheless water may be obtained in them by wells of smaller depth than are required in the main part of the upland.

The origin of the Cotswold dry valleys has been variously explained. It is certain that they have been cut by water and in most cases the problem is to explain the disappearance of the stream. It has been suggested that many of these dry valleys were cut by meltwaters flowing from snow or ice caps which were situated on the higher uplands during the Ice Age. This part of England was not covered by glaciers, as were areas farther north, but during the more intensely cold periods snow may have accumulated on the higher ground, giving rise when it melted to temporary but powerful streams. This explanation of the drying of some of these valleys is supported by the fact that they 'hang' above the stream-bearing valleys to which they are tributary; the main valleys, watered from springs which have continued to flow, have been progressively deepened since the ice disappeared, leaving the springless tributary valleys in disharmonious junction with the main stream. Dry valleys also occur in the Chalk, and these are discussed more fully in the chapter on the chalklands. Some of the arguments put forward there may also be applicable to these Cotswold valleys, but the Chalk is a porous limestone with very different properties from those of the oolitic limestones.

Within this upland region there are no large towns: on its flanks lie Bath and Gloucester to the west, and Chippenham and Swindon to the east, but these belong rather to the vales than to the upland, except in so far as they have formed markets for the exchange of the produce of the adjoining regions with their differing products. The towns of the Cotswolds themselves are smaller and most of them have grown very little for many years. Few are situated in the highest part of the upland, Minchinhampton and Stow-on-the-Wold being almost alone in their hill-top situations. The other Cotswold market towns have lower sites, usually where old routes cross the valleys. Cirencester, Tetbury, Malmesbury, Northleach, Chipping Norton and others are spaced at intervals of ten or twenty miles, that is at such a distance that market centres were within suitable reach of all parts in horse-transport days.

Many of these warm, brown towns were more important in medieval times, when they were centres of the wool industry; then the value of Cotswold fleeces was widely recognized, and from these market towns wool was exported to the Continent. In many cases the towns were beautified by the merchants who prospered during the fourteenth century, and who built or extended churches and wool-markets, and they have been little changed in recent years, although their importance declined when the wool was no longer exported for weaving.

Many of them are spacious towns: their streets are wide, for the main street has served as a market-place. The better houses are built from large well-dressed stones, with windows of fine masonry. Many have steeply pitched gabled roofs formed of Cotswold slates (Plate 2). The churches and more important buildings have derived their stones from well-known beds in large quarries: some stones of Malmesbury and Lacock Abbeys were from Box. But the materials for smaller buildings were not carried for any great distances, and they were usually dug in the neighbourhood of the town itself. Even in the smallest cottages, however, the stonework of the chimneys and of the windows is carefully dressed, though it may be placed side by side with rough walling. Some of the towns which are built on flatter ground, so that the view does not include a massing of buildings, are much less pleasing; this is the case at Marshfield, which has some fine houses but with the absence of trees is one of the dreariest towns on the hills.

The villages of this area share these same characteristics as regards buildings. But they show greater variety in arrangement and in scenic features. Many of the cottages date from the sixteenth century but few have conspicuous architectural dating.

The Valleys of the Cotswold Edge

The towns and villages of some valleys nearer the Cotswold edge have more individual characteristics. They are situated in valleys which are far deeper and narrower than those of the main part of the upland. This results in part from the fact that the rivers of this area are flowing along a steep path to the sea in the Bristol Channel rather than into the Thames, and this has enabled them to cut down their beds more rapidly. The two most important of these west-flowing streams are the Bristol Avon and that Frome which traverses the Stroud valley (for each of these names is shared by several rivers). These two rivers, with some smaller streams in the same area, and with their various tributaries, have cut valleys which extend for miles into the upland area, and these features are thus confined to the areas around Stroud and Bath.

The Avon rises above Malmesbury as a stream essentially similar to the dip streams of the upland area, but shortly it turns almost at right angles, to follow a wide open valley through Chippenham (where a narrowing of the valley makes possible an easy crossing) to the picturesque town of Bradford-on-Avon, whence by a deep gorge the river winds its way through the oolite uplands to Bath. The rest of its course to the sea is equally anomalous, and is described in the Bristol and Somerset chapter (pp. 323–5). The Frome, while it does not commence as a dip stream, intercepts such streams above Miserden, and then likewise flows south to Sapperton where it also turns sharply west and cuts a deep narrow valley from which it escapes at Stroud.

The cutting of deep gorges by these streams has produced here quite a different type of relief compared with that of the main part of the uplands. The upland surface is here more broken up or dissected than farther east, for whereas in that area the valleys often form mere channels in a moderately even surface, the valleys here cut the surface into narrow and irregular strips, pieces of the upland surface being frequently cut off from the main mass.

Examples are well seen in Cam Long Down, which when seen from the hill above Dursley is clearly recognizable as a detached portion of the neighbouring upland which has been isolated by the efforts of the little stream flowing into the Dursley valley. The long irregular tongue of upland culminating in Stinchcombe Hill is only connected with the main area of high ground by a ridge a few hundred yards wide between Waterley Bottom and the Dursley valley; it is interesting to notice how the only good road on to this tract makes use of the 'isthmus'. Obviously a little further erosion by these streams, carrying their sources farther back into the plateau, will separate this also as an outlier. From a consideration of these instances, it is apparent that the rivers have produced much of the present topography, and an examination of this area shows also how such outlying hills as Robin's Wood Hill and Dundry have been cut off from the Cotswolds.

It is, however, by no means easy to explain the anomalous courses of these streams. It is clear that streams flowing down the scarp face, from near the scarp summit to the plain a short distance away, must have a very steep gradient and must therefore flow almost as torrents: their average gradient is in many cases as much as one in ten. They descend to 200 feet above sea-level after flowing possibly a mile, whereas the streams flowing on the dip slope may flow some scores of miles before they reach an equivalent level. Thus the scarp streams have much greater cutting power than the dip streams, and they not only deepen their valleys but simultaneously carry back their sources, extending their valleys into the upland.

This tendency of swiftly flowing streams to push back their sources into the upland frequently results in such a stream intercepting the drainage of another valley. Such extension of west-flowing streams and capture of dip streams may perhaps explain the development of the west-flowing portions of the Frome and the Avon.

These valleys around Stroud and Bath have other peculiarities. Their towns and villages are hemmed in by the steep hills, and houses are ranged in picturesque fashion along the tree-clad slopes. The Avon valley about Limpley Stoke, just above Bath, is well known: the Frome valley (better known as the Stroud valley and aptly named the Golden Valley) is no less attractive; it is glimpsed from the railway line into Gloucester, but it merits closer inspection. The tributary valleys are more beautiful than those of the main streams, just as their villages are more unspoiled than such towns as Stroud, which shows too much modern expansion to allow comparison with the older Cotswold towns, or even with Bath. This latter is best admired from a distance, as from the road running north along the Cotswold edge or from Combe Down: then the creamy yellow of its terraces, its beautiful bridges, and the sombre Abbey emphasize the delight of its situation in the hollow of the hills.

The valleys tributary to the Avon both on the north and south have been deepened for many miles. One of the most interesting is the By Brook which joins the Avon at Bathford after intercepting numerous dip streams on its right bank. So deep is the valley carved by this brook that the Fosse Way from Bath to Cirencester was kept to the west of it and turned abruptly at its head. The most interesting village in this valley is Castle Combe, perhaps the most perfect village of the Cotswold belt. The combination of its reputation and the number of visiting cars may have made its beauty a little tawdry, but every cottage is a picture, every view is delightful. The brook runs for some way through the village street, and the irregular grouping of the houses around the market cross, with the thickly wooded hill slopes, makes a memorable scene. Farther down the same valley is the smaller and similar but less-known village of Slaughterford, while in close proximity are such upland villages as Biddestone, which differ in having had greater room to grow, and spread themselves casually around the extensive village greens (see also Plate 2).

In the valleys of the Stroud area are many larger villages, which

Plate 2. *Lower Slaughter, Gloucestershire, an example of the use of oolitic rocks as building stone. In this area of plentiful and suitable stone, walls, roofs and also bridges can be built from it. Contrast this with Plate 17.*

grew to importance with the growth of the home woollen industry. The waters of this area proved well suited to dyeing, and hand looms were formerly found in nearly every cottage. Later the mills used water power, and small modern factories are now to be found in most of these valleys, but they do not detract greatly from their picturesqueness. This area became the great broad-cloth region, while towns of the uplands, handicapped in many cases by lack of water (and of water power), lost their wool trade, or, as in the case of Tetbury, took up such branches of the industry as wool-combing and spinning. But many of these little towns, though situated in the deep hollows within the Cotswold belt, are rather of the vale than of the upland, and places like Nailsworth show a mixture of building materials, brick and tile being used side by side with stone in a way quite unlike that of the real stone area, but here without any loss of attractiveness.

Southwards from Bath and Frome, the geological map becomes more complicated, and the stone belt forms a much less impressive feature, but it may be traced through Somerset and Dorset, forming wooded scarps rarely over 500 feet high. Several distinct ridges are present in places, running more or less parallel to one another. The most prominent overlooks the Somerset plain near Bruton and Castle Cary, extending south to beyond Sherborne. To the east these ridges slope down into the clay country, in part of the Vale of Blackmoor so well described in Hardy's *Tess of the d'Urbervilles*. The stone belt reaches the south coast near Bridport, where a thin layer of oolitic rock caps the cliffs. These areas, and also the Cotswolds, will be seen on the geological map printed on the back cover. They are the Middle Jurassic rocks, that is the Oolites and the Cornbrash.

The reason for this change in the character of the stone belt south of Bath must be sought in the conditions under which the rocks were deposited. The rocks laid down at the same time as those of the Cotswold area stretch from Dorset to Yorkshire, but they differ considerably in character over this length. They are all marine rocks, but the sea in which they were deposited was in four almost distinct basins, separated by ridges running from east to west or from south-east to north-west across it, covered only by shallow water, and thus sediments differing in thickness and character were accumulated in each of the basins. One of these former ridges separates the Cotswolds from the Somerset and Dorset stone belt, and its line can be easily recognized today as a continuation of the Mendip Hills, which are older than the Cotswolds. Another old axis separates the Cotswolds at the other end from the Northamptonshire uplands to the north-east, in the area of the Vale of Moreton and Oxford.

The Clay Country East of the Cotswolds

All along its extent the stone belt sinks gently to the east or south-east. As the limestones dip more steeply than the slope of the land surface (Fig. 1), they eventually pass beneath the clays flooring the vale which stretches, with small intermediate elevations, almost to the foot of the chalk hills away to the south-east. Where the Cotswolds join this belt of clays a line of towns and settlements occurs, some prospering, as we have noted, as market or exchange towns between the adjacent regions, most of them benefiting by the more abundant water supply which becomes available when the clay belt is reached. Along this border-line more varied agriculture is possible, for whereas the limestone soils are light and dry, some soils of the adjoining region are heavy. On these clays the beech woods of the limestone tract are replaced by denser woods of oak, while great areas are devoted to meadowland, the chief activities being cattle rearing and dairy farming.

This clay belt stretches through Trowbridge to Oxford and Bedford. It is interrupted by a discontinuous ridge which is well seen in the neighbourhood of Oxford, where Corallian limestones (often similar in general character to the coral limestones of the Cotswolds, but of somewhat more recent age) give rise to an escarpment quite comparable to that of the Cotswolds but of smaller elevation, the position of which is marked on Figure 2. This ridge, with its concave scarp facing to the north-west and its dip slope to the south-east, rarely exceeds 300 feet in height, but in this wide, flat vale it makes a conspicuous line of hills, especially east of Faringdon, which are rather too emphatically known as the Oxford Heights. To the north-west of this the clays are known as the Oxford Clay, to the south-east as the Kimmeridge Clay, but both are grey-blue clays and for our present purpose further distinction is unnecessary. These, with the Corallian, are the Upper Jurassic rocks marked on the geological map on the back cover.

In the clays the rivers are able to cut down their beds more rapidly than on the limestones: as these outcrops are lowered they tend to form valleys and the dip streams from the Cotswold slope frequently turn at right angles to flow along the clay belt. They thus follow the strike of the rocks, as in the case of the Thames in the Oxford Clay vale west of Oxford. Subsequently the Thames turns once more to follow the dip and cuts through the little escarpment of Corallian limestone. The situation of Oxford was determined by this gap in the miniature ridge, which provided a dry-land route in a clay district where formerly there were extensive marshes (Fig. 2).

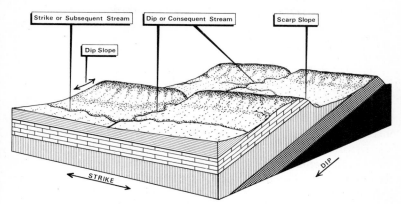

Fig. 3. *Scarp and vale scenery and its drainage.*

The tendency of rivers in this region to form systems with a more or less rectangular pattern illustrates a fundamental feature in river evolution. Dip streams or consequents determined by the inclination of the original slope on which drainage was initiated represent the first stage in river development. Where softer beds occur, however, tributaries presently begin to develop at right angles to them along the outcrop of the softer rocks, and where rapid erosion is possible, these streams (known as subsequents because they are only formed after the dip streams, or as strike streams because they follow the trend or strike of the beds) become the more important in the river system. But for a long time the pattern of tributaries at right angles to the main streams persists, and in country of simple structure controlled by gently dipping beds as in the case of the Cotswolds, this 'trellis' pattern is almost always characteristic (Figs. 2 and 3).

In the clay belt, stone has been used only for churches and larger houses, and of course for the Oxford colleges: for almost all of these the Cotswold quarries have provided the materials. But the city of grey towers is rapidly becoming a city of red roofs, and the cottages in this tract are mostly of brick: there are extensive clay pits in many areas, especially near Bedford and Peterborough. This area of Oxford Clay produced a high proportion of England's 'railway bricks', and they are still produced in vast quantities for all kinds of building, near and far from the brickyards. So that we find in travelling from the stone belt towards the east we pass from towns that are almost wholly of stone, such as Bradford-on-Avon, to Trowbridge, where a moderate proportion of stone houses is still found but side by side with red-brick houses, to Westbury where stone is rare, as it is also in most of the villages in the clay area. The suddenness of the change in domestic building materials in passing from one area to another suitably emphasizes the local source of most of the materials; equally noticeable in its scenic effect is the replacement of stone walls by hedges. A fresh change in techniques and use of materials results in the extensive quarrying, for concrete making, of gravels found locally in the upper Thames valley, leaving a large number of 'wet' pits near Cirencester and Fairford, some of which are now being landscaped to form the 'Cotswold Water Park'.

3. The Northern Extension of the Stone Belt

Following the Cotswold belt to the north-east, the Northampton uplands and their extension into Leicestershire next call for comment. Essentially the structure is quite similar to that of the Cotswold area. From the crest of a steep escarpment facing westwards there is a wide view across the plain of the English Midlands, notably from Edge Hill, while to the south-east the surface slopes gently down to the clay plain, its dip face showing other smaller scarps owing to successive hard bands coming to the surface. The north-western face of the ridge in this area is made by a group of ironstones, the Middle Lias ironstone or Marlstone: the equivalent of this rock was mentioned in describing the Dursley area, where it forms a low platform or terrace along the face of the main scarp. Northwards this rock becomes thicker and stronger and itself forms the front of the upland for many miles. The oolitic limestones form more or less distinct and parallel minor scarps. This upland has less altitude than that of the Cotswolds, its height being little over 500 feet, but it has been deeply dissected by several rivers, as by the Cherwell and its tributaries at Banbury and by the Nene at Northampton. These valleys compare in character with those of the Stroud area but they are generally wider and more open. They are important in the communications pattern, one gap near Northampton containing canal, railway, road and now the M1.

From this description it can be seen that the rocks differ slightly in character from those of the Cotswolds, the Marlstone being here the principal scarp-forming rock, as in Edge Hill, while it is quarried for iron at other places in this tract. The oolitic limestones, so prominent in the Cotswolds, are here not so important in their effect on the scenery. This is because the oolites of the Cotswolds were laid down in one former basin, and the corresponding rocks of Lincolnshire and Northamptonshire in another, separated by a transverse ridge between Moreton-in-Marsh and Oxford, which was covered only with shallow water.

This area also has its famous building stones, the quarries at Barnack in north Northamptonshire having yielded stones which were in constant use in medieval England, supplying stones for Peterborough and Boston and for many churches in the Fen country, and for other areas where they could be carried by water. Ketton in Rutland likewise supplied the stone used not only locally but in some of the Cambridge colleges. Stamford is a most delightful town of stone buildings. But a far greater extractive industry here is the quarrying of the ironstone itself, providing ore to support the steel works at Corby.

On through Leicestershire a similar type of country continues, the Middle Lias ironstone forming the sharp ridge on which Belvoir Castle stands, overlooking the Vale of Belvoir on its north-west. Through much of its outcrop this ironstone produces chestnut-brown or orange-coloured soils, and imparts a richness to the colour of the fields; it has been used in many of the walls, which appear quite distinct from those of the oolite tracts. The ironstone is extensively quarried at many places along its outcrop. This is a bedded ore quite different from many iron ores, which occupy irregular veins traversing other rocks; the ironstone of workable quality here forms a bed some ten feet or more in thickness, which extends for many miles along the outcrop and south-eastwards passes underneath the newer rocks which occur above it. The quarries are shallow and extensive, the thin soil being removed to an area already worked as the ore is quarried, the fields from which the ore is dug being thus lowered by some ten or more feet below the level of the roads or of any unworkable areas, and the country thus presents a curiously uneven appearance.

The Lincolnshire Stone Belt

The Lincoln Cliff is the most striking feature in the county. Running almost due north and south it rises very steeply on the west from the extension of the Midland plain. It has no great elevation, for it rarely rises more than 150 feet above the low ground to the west, but its straightness and sharpness make up for its lack of height, and facing such a wide expanse of flat and often marshy country, its scarp is no less impressive than that of the Cotswolds. East of the edge is a dip slope some four or five miles in width, known as the Heath, an area which is commonly treeless and almost devoid of surface streams: it has a few dry, shallow valleys, and in this feature, and in its dry-stone walls, it greatly resembles portions of the Cotswold stone belt. For long, however, it remained almost uncultivated, and much of it was heathland until the Napoleonic wars: as a consequence it has few villages and it lacks much of the attractiveness of the Cotswolds.

East of the Heath is a wide clay vale, and as is the case farther south, where the limestone tract meets the clay there is a chain of settlements. Here also the clay has been readily eroded, and several streams follow the outcrop in strike valleys, the Ancholme to the Humber, and the Langworth, rising near the Ancholme within the clay vale itself, to the Witham and so to the Wash. This is a poorly drained area of fenland, with an indeterminate watershed between the rivers, and some artificial straightening of channels. It is part of the bed of a lake which existed for a while during a late stage of the Ice Age, when the Humber and Wash drainage exits were blocked by ice (see Chapter 10, p. 165, and Fig. 17).

Along the western face of the Lincoln Cliff are numerous small villages, their sites determined by the spring line at the junction of the limestone and the underlying clays, for much of the rainfall of the limestone area passes underground to flow out where the clays prevent further downward percolation. These villages are at almost regular intervals of one mile and, as in the case of Coleby and Boothby Graffoe, their parish boundaries extend for a greater distance east and west than from north to south: in this way each parish includes below the village an area of low meadowland, and above it a portion of the drier Heath, an arrangement which is frequently found along the junction of two areas of different type. The low land beneath the western face of the cliff was also part of a lake bed at some time in the Ice Age (Fig. 17).

As has already been pointed out, the most impressive feature of the cliff is its simplicity and regularity. From just north of Grantham to the Humber the scarp forms an almost unbroken line, scarcely crossed by any rivers. In about fifty miles it is only cut by one gap of any importance, that at Lincoln. Here the Witham has cut through the cliff a narrow gorge, on the north of which stands Lincoln, with the castle guarding the crossing of the river by the old route which naturally followed the crest of the scarp, above the swampy lowlands. The cathedral now dominates the city, and the view from the west, with the old town clustered along the steep hill slopes above the river, is one of the happiest in England. But like other similar gaps, whether through escarpments or mountain ranges, the Lincoln Gap has also affected more recent transport development, save that the railway uses the gap to escape the high ground whereas the old roads were forced to descend at the gap: the result, however, is similar, for Lincoln has extended rather awkwardly over the plain in response to its development as a railway and engineering centre.

The Lincoln Gap affords a clear illustration of the tendency for rivers to follow the strike of the soft beds rather than to flow with the dip, which involves crossing the outcrops of soft and hard beds alike. For formerly much of the Trent drainage which now reaches the sea by the Humber flowed through the Lincoln Gap, there being still an almost open path for Trent water along this course, through which even in recent times flood water from the Trent has found its way into the Wash. But the lower Trent has cut down its valley along the soft clays to the Humber and has left only a diminutive Witham in possession of its original gap.

To the south, a gap at a higher level at Ancaster, now carrying no through drainage, marks an even earlier course of the Trent as a dip stream. The course of the Trent as a whole is discussed more fully in Chapter 10 (pp. 162–5).

This area cannot be left without commenting again on the quarrying of the Lias ironstones, leading to the establishment of an iron and steel industry near Scunthorpe.

The Moors of North-East Yorkshire

In Yorkshire the stone belt is only represented in a very modified form. The moors of north-east Yorkshire may be regarded as its continuation, but they differ greatly in many ways. For much of this extensive upland tract, some thirty miles wide and fifteen miles from north to south, is beautiful heather moorland over 1,000 feet high, cut by deep, steep-sided dales. Generally the up-land surface is smooth or gently undulating, and the dales are narrow. The chief roads therefore avoid the valleys and afford wonderful views over the moors; those from Whitby to Pickering and Scarborough are well known for their beautiful scenery.

Limestones are here much thinner than in the Cotswolds or even in Lincolnshire, and these moors are made up of a varied series of rocks: shales, ironstones, grits and limestones are all present, but most of the moors are made up of resistant deltaic sandstones of the same age as the Cotswold oolites. The region thus lacks some of the essential characters of the simpler parts of the stone belt, and there is more variety of soil and land form. There are no such characteristically limestone soils as in the Cotswolds, and while grey stone is much used in building, red tiles are also very abundant.

This change in rock character is again due to the fact that the rocks were deposited in a structural basin distinct from that of Northamptonshire and Lincolnshire. The two basins were separated by a ridge in the neighbourhood of Market Weighton, about ten miles north of the Humber, where the Jurassic rocks are now very thin, so thin that they are overlapped by the younger Chalk in places. The northern basin of deposition was much deeper and so sediments are thicker. The reader who is following the chapters consecutively will now have read of the four struc-tural basins, Somerset–Dorset, Cotswolds, Northamptonshire–Lincolnshire, and Yorkshire, separated by the three old ridges of Mendip, Moreton–Oxford, and Market Weighton.

The highest parts of the north Yorkshire moors are mostly to be found along the east–west line running from Goathland through Glaisdale Moor and Westerdale Moor; from this line the dales generally run north and south, the streams flowing north being tributary to the Esk while those going south enter the Vale of Pickering.

In view of this arrangement of the streams it is not surprising that they have cut no through route across the area: each valley normally ends up among the moors, though Bilsdale extends so far to the north that it almost reaches the scarp edge, and provides one valley route across the area.

Within the moors themselves there are no towns and few villages, for on account of the elevated position, little ground is cultivated. Many villages are to be found around the borders of the area, however, and especially along the southern edge, where the dales open out into the Vale of Pickering; here are situated the old market towns of Pickering and Helmsley, and a dozen smaller places. The dales, however, have always been quiet, and Rievaulx Abbey in Rye Dale, not far from Helmsley, has a very lonely site, hemmed in between the river and green hills, scarcely rivalled in its beauty of setting by any ecclesiastical ruin except possibly that of Tintern.

The northern and western borders of the moors are very distinct. On the west they rise sharply from the Vale of Mowbray where the scarp, terraced owing to the harder bands among the shales, rises in a few steps to over 1,200 feet in the Hambledon Hills. This scarp is somewhat fretted, but only by swift scarp streams flowing into the vale. Farther to the north-west and north there are some outliers of the moors beyond Guisborough Moor, of which Roseberry Topping is perhaps most conspicuous, capped by a bed of hard grit, with terraced slopes owing to the alternation of harder and softer beds. In places the grits have weathered into strange forms; the Bride Stones near Pickering are best known among these.

Along the northern border of the moors, that is, particularly along the Cleveland Hills, the ironstone bands of the Middle Lias, which are mined in Northamptonshire, are again of workable quality. Although they were known and dug much more than a century ago, it was only about 1850 that they led to the development of the important iron industry which now centres on

Plate 3. *Newtondale, a major glacial meltwater channel. This meandering valley, some 250 feet deep at this point, is at present drained by a stream so small that it cannot be seen in the photograph. One must, however, imagine torrents of glacial meltwater cutting this valley at a late stage of the Ice Age. The rocks here are sandstones and shales, with a thin band of ferruginous limestone near the top of the slopes.*

Middlesbrough, a town which grew from a population of 154 in 1831 to over 7,000 in 1851, and is now part of the Teesside industrial area. While the iron and steel industry continues on a large scale, it now uses imported ore, and the Cleveland ironstones are no longer mined, the last working being in 1963; a fine example of geographical inertia.

We must now turn our attention to Newtondale, the only continuous channel across the moors from north to south (Plate 3).

It was followed by the former railway from Pickering to Whitby, while the main road follows a more direct path over the moors just to the east. For the greater part of its length this is not a river valley, for water drains both north and south from an indeterminate watershed in Fen Bog. The bog itself shows that the small supply of water draining from Newtondale has inadequate drainage, so that peat bog develops. Newtondale has steep sides and a wide and nearly flat floor: unlike the other dales which are generally wider on the softer rocks, its form is little affected by the rocks which it traverses. These facts suggest an old glacial spillway or overflow channel, and it is to conditions in the Ice Age that we must turn for an explanation of Newtondale's anomalies. During this period, comparatively recently in Britain's geological history (see Chapter 22), ice sheets covered much of the country almost to the Severn–Thames line, in a number of separate advances. The more recent of these did not reach so far, but being more recent, have left more definite proof of their presence to be seen in the present landscape.

At a late stage in the glaciation we may imagine ice nearly surrounding the Cleveland Hills and the moors; indeed it is thought by some that the highest parts of the north Yorkshire Moors were never overrun by ice at all, and remained as an isolated island surrounded by ice at the maximum glaciation. Later than this we can imagine the ice surrounding the moors and extending south in the Vale of Mowbray and as far as the moraine now well seen at Escrick, just south of York; to the east the ice sheet covered what is now the coast at Whitby and between Scarborough and Filey (Fig. 4). Thus the present mouth of the Esk at Whitby was blocked by ice, and it was thought until recently that the ice thus caused lakes to develop in the dales. The waters rose in these lakes, held up by the ice and the hills, until eventually the water poured over the moors at their lowest point into the Vale of Pickering, the rush of water carving Newtondale. Recently, however, the evidence for the lake or lakes in Eskdale has been re-assessed and questioned, and it is now thought that ice overran the moors to the north of Eskdale and penetrated down the dales. However, it remains certain that Newtondale was carved by glacial meltwater, whether from a lake or from beneath ice. The amount of this meltwater can be imagined from the size of the valley, and of the delta deposited by the waters when they left the hills, the delta on which the town of Pickering now stands.

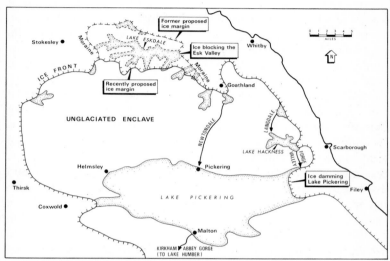

Fig. 4. *Glacial features of north Yorkshire.*

The Vale of Pickering

At this stage some reference may appropriately be made to the Vale of Pickering, which lies on the southern border of the moors and which extends for some thirty miles from the coast at Scarborough and Filey as far west as Helmsley. The vale is shut in between the moors on the north and the wolds on the south, and immediately impresses by its remarkable flatness. Writing of it nearly 200 years ago in his *Rural Economy of Yorkshire*, Marshal said that 'Nature, perhaps, never went so near to form a lake without finishing the design', but in fact the lake had been formed and the vale represents a lake which has been drained rather than one incompletely formed.

The unusual nature of the vale is strikingly shown by the amazing course of the river Derwent which drains it. Rising in the moors
not far from the coast, and joined by the Hertford River which rises
almost on the shores of Filey Bay, it flows for many miles westwards, away from the coast, turning south near New Malton to
leave the vale by a path through the Howardian Hills in a gorge
some 200 feet deep, yet so narrow that it seems a small landslip
might block it. The explanation of this anomalous course is to be
found in the coastal area near Scarborough and Filey, where
mounds of glacial deposits tell of the ice sheet which filled the
North Sea and impinged on the coast (bringing with it material
from Scandinavia whose source can be traced with such certainty
that the direction of the ice flow is beyond dispute); as well as
damming the Esk valley, the ice held up the drainage which until
then had passed from west to east along the Vale of Pickering. The
dip streams from the moors at that time joined a strike stream
flowing east along the wide outcrop of the Kimmeridge Clay which
underlies the vale, and a wide area of low relief had already been
carved out in the soft rocks. The ice dam at the valley mouth led to
the flooding of this valley and the formation of a wide lake, into
which the streams from the dales continued to flow, joined by the
waters from the Esk valley, via the meltwater channel of Newtondale, as already indicated. With this abundant inflow of water
the floods rapidly rose and the lake expanded until the waters
reached the level of the lowest available outlet: this happened to
be on the southern edge, at the position of the present Kirkham
Gorge, south of New Malton, other areas where the land was
lower being then blocked by ice (Fig. 4). Thus an important spill-
way was established, carrying the meltwaters from the whole
Cleveland area into the Ouse basin to the south, itself then the site
of a rather similar lake, caused by the ice sheet blocking the mouth
of the Humber (Fig. 17). The Ancholme–Witham area described
above was also an arm of this former Lake Humber. Naturally such
a vast flow of water possessed considerable erosive powers, and the
shallow col by which it at first escaped was deepened to form the
deep gorge in which Kirkham Abbey is situated. With the deepening of the gorge the level of the water in the lake was lowered, and

eventually the whole was drained, but not before the floor had been covered by wide layers of clay and silt, and the deltaic deposits had accumulated at the mouths of the dale rivers. With the disappearance of the ice it might have been expected that the drainage would be re-established in something of its original condition, but the mounds of boulder clay left near the seaward end of the vale were just high enough to prevent the flow of streams to the coast.

So the peculiar characters of the Vale of Pickering may be explained. The present Derwent, flowing for so many miles with so little slope, is exceedingly sluggish, and this, together with the nature of the alluvial deposits over most of its area, has made much of the vale a marshy area, parts of which have only been drained in recent years. In the centre of the vale are damp cattle pastures of richest green, but around its fringes, and especially where the ancient deltas have provided slightly higher and drier ground, there is some cultivation. Most of the towns and villages, as we have seen, are to be found along this fringe, but small and low 'islands' within the marsh have provided sites for Kirkby Misperton, Edston and Normanby.

The Yorkshire Coast

The Yorkshire coast stretches from the Humber in the south to the Tees in the north. The southern part of this coast, consisting of the chalk cliffs of Flamborough Head and the low boulder-clay cliffs of Holderness, is included in the next chapter on the chalklands. The coast of the Yorkshire moors area, which is described here, exhibits very clearly its main structures. The high cliffs show a sequence of almost horizontal or slightly undulating beds, with grey shales and clays, or brown sandstone and limestones predominating in different parts. Owing to the strong joints which cross the beds almost at right-angles, and along which the rocks break away under the action of the waves, the cliffs are often nearly vertical, but the clays tend to wear down into smoother slopes more quickly than the sandstones and limestones, so that the variation of rock types introduces a fascinating diversity of cliff form and colour (Plate 4). The changing dip brings first one bed then

Plate 4. *Whitby harbour, Yorkshire. This narrow mouth of the river Esk has been in existence only since the Ice Age, during which boulder clay blocked the former mouth near Upgang. A fault is found roughly aligned with the harbour, and the beds in the cliffs on the far side of the harbour are some 200 feet higher than their counterparts on the near (west) side. The rocks in the cliffs, bedded nearly horizontally, are mostly shales and sandstones, locally known as 'Dogger', with Alum Shale below.*

another to form the curved reefs which are exposed on the foreshore at low water. Where the softer clays of the lower part of the Lias are brought up into the cliff by gentle folds the coast has been cut back to form bays, such as Robin Hood's Bay. And in these are placed delightful villages of red-tiled cottages which are perched irregularly on the steep hillsides. Around the coast there are in fact more settlements than in the moors themselves, and Whitby and Scarborough are the largest towns in the area, although these have a life which is almost independent of the moors.

At intervals along this coast quite another element enters into the construction of the cliffs, and produces a strongly contrasted type of coast scenery. This consists of the deposits formed by the glaciers to which reference was made above (Fig. 4). As these melted they left behind a vast irregular mass of material that they had gathered from the rocks over which they had passed, and which included boulders of all sizes and many kinds, together with finer material resulting partly from the grinding together of these larger blocks: this mixture of coarse and fine debris was dumped in unsorted heaps showing no trace of bedding when the ice melted away. This occurred in the valley of the Esk, which, before the Ice Age, entered the sea some way to the west of Whitby, in the neighbourhood of Upgang. Here a deposit of 'boulder clay' completely blocked the original mouth, forcing the river to follow a lower outlet which lay over solid rocks. Through these it has cut the gorge at whose mouth Whitby stands, and thus on both sides of Whitby the cliffs are made of regularly stratified rocks of Jurassic age (Plate 4), whereas at Upgang the cliff consists of an unstratified structureless mass of boulder clay which presents a deeply channelled and exceedingly irregular and unattractive surface. At other places along the coast, notably near Filey, similar cliffs of boulder clay are found.

Inland, boulder clay also builds up much of the low ground, especially near the coast and in the Esk valley and its tributary dales. Egton Low Moor, west of Whitby, is based on such clay deposits.

4. The Chalklands

The chalk areas of England, which Huxley considered so suggestive of mutton and pleasantness, are popularly thought of as wide expanses of grassy downland and hills with smooth rounded curves. In the mind's eye, the short grass barely covers the white rock, for the soil is extremely thin, and the white gashes of chalk pits on the hillsides and the pale cream of the rough flinty tracks give to these areas a lightness, a delicacy of colouring, which is unlike that of any other hill country. The wide extent of such a chalkland view is best seen in the clear lights between the broken storm clouds, when the swiftly moving shadows throw into relief the contours of the hills.

On these rolling hills clumps of trees crown many summits, while great areas were formerly given up to sheep pastures. Nowadays, with changes in farming methods, sheep are comparatively rare, and there are wide stretches of open tillage. Farms are few, and are so placed in the valleys that in a view over miles of country scarcely any are in sight. Most villages are dotted along the valley bottoms, while such few towns as occur in the chalk tracts are to be found on water courses: of these Salisbury is a notable example, being situated near the confluence of several rivers.

This is an idealized picture, however, and is typical only of areas of bare Chalk carrying little or no superficial cover, notably Salisbury Plain and parts of the South Downs (Plate 5). Other parts of the extensive chalk outcrops have a cover of various superficial deposits, and have had a different history, and so vary considerably from this ideal picture.

The present chalk outcrops can be visualized as radiating from Salisbury Plain, and cover wide areas of England (see geological map, back cover). The largest branch lies north-eastwards, comprising

Plate 5. *The South Downs near Fulking, Sussex; typical chalk country with no super-ficial cover and very few trees. The indented nature of the chalk scarp is seen well, also the dry valley systems on the dip slope. The dip slope (on the left) is of Upper Chalk, while the scarp slope is in Middle Chalk. The top and bottom of the scarp are followed respec-tively by hard bands in the Chalk, the Chalk Rock and the Melbourn Rock. The Lower Chalk extends to the right of the scarp slope almost as far as the road, while the villages of Fulking and Edburton and the road between are on the narrow outcrop of Upper Greensand. To the right of this is the Gault Clay. In the background is one of the gaps through the Downs, that of the river Adur. The flood plain of the river can be picked out. Compare with Plate 6.*

the Marlborough Downs, the Berkshire Downs, the Chilterns. East Anglia and the Lincolnshire and Yorkshire Wolds. This branch now has a south-easterly and easterly dip, and thus for most of its length is parallel to, and has a scarp slope or slopes looking out over, the Jurassic rocks described in the previous two chapters. Part of the Chilterns and all the chalklands to the north have been overridden by ice at some stage of the Ice Age, altering the appearance of this country: it rarely, if anywhere, conforms with the ideal picture.

Another branch radiates south-west from Salisbury Plain, through Wiltshire and Dorset, with a scarp of varying impressiveness facing westwards. Beyond this main scarp, small areas of nearly horizontal Chalk stretch into Somerset and just into east Devon. A finger of Chalk runs through south Dorset to Swanage, and is continued in the Isle of Wight.

Eastwards from Salisbury Plain run two branches, to the north the Hog's Back and North Downs, to the south the Chalk of eastern Hampshire and the South Downs. The North and South Downs form scarps which face each other across the older rocks of the Weald, and so it is possible to deduce the present structure of the chalk outcrop (Fig. 5). Everywhere along the main western scarp of the Chalk one looks out over older rocks which pass beneath the Chalk, except in the Fenland area. If one descends the dip slope of this chalk escarpment, one comes to younger rocks, the Chalk passing beneath them. The largest exposures of these younger rocks in England are in the two basins of London and Hampshire. Beyond these basins the Chalk rises again, in the case of the London basin in the North Downs, and beyond the Hampshire basin in Dorset and the Isle of Wight. The Weald is an area of older rocks between the North and South Downs which has been upfolded.

The chalk surface has thus been folded, although we may surmise that the Chalk was laid down in nearly level sheets of wide extent. It is a soft white limestone of marine origin, and is often of great purity, being made up almost entirely of calcium carbonate derived from shells. Small shells of Foraminifera are present, together with fragments of larger shells, and a high proportion of very fine-grained calcium carbonate, probably derived from the

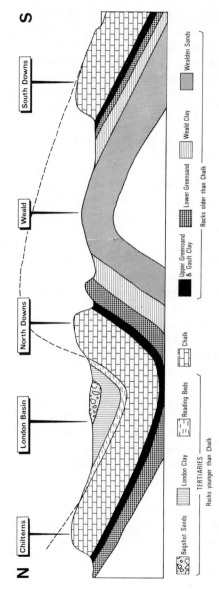

N

S

Chilterns | London Basin | North Downs | Weald | South Downs

TERTIARIES
Rocks younger than Chalk

Bagshot Sands | London Clay | Reading Beds

Chalk

Upper Greensand & Gault Clay | Lower Greensand | Weald Clay | Wealden Sands

Rocks older than Chalk

Fig. 5. *Sketch section through the London basin and the Weald, showing the major structures. Note the asymmetry of the folds.*

disintegration of other large shells. Although the Chalk appears to be uniform in character, some beds are harder than others, while the lower part in some places is particularly marly and soft.

Wide sheets of Chalk were deposited over much of what is now England, and possibly also Wales. The deposits were laid down in a sea which transgressed wide areas, following deposition of older rocks under different conditions. The rocks described in the previous two chapters were laid down in their four nearly separate basins in a period known to geologists as the Jurassic. The lower part of the succeeding Cretaceous period is represented also by varying rocks; the lower beds of the Weald were laid down in a lake, while later marine beds were laid down in two basins separated by an axis from north-west to south-east through the area of ancient rocks of Charnwood Forest. Scenically the southern deposits are more important: these consist of the Upper and Lower Greensand, separated by the Gault Clay. Conditions in the southern basin were far from uniform, however, and in places some of these rocks are missing from the succession. Following this varied deposition the extensive transgression of the chalk sea took place, and more uniform conditions of deposition obtained.

It is important in understanding present chalk scenery to trace briefly the subsequent history of the Chalk after its deposition. The north and west of Britain were uplifted, and the chalk sea retreated. Results of this uplift include the regional dip to the east and south-east contributing to the development of the English scarplands, and a north-west to south-east drainage system, remnants of which still survive. An example of this is the lower course of the Bristol Avon, noted in the Cotswolds chapter, and others will be mentioned presently (p. 61). Although the direction of flow of the Avon has now been reversed to the Bristol Channel, its discordant character may perhaps be explained by superimposition from this former chalk cover.

Before the next beds in the depositional record were laid down, the folding which resulted in the London and Hampshire basins and the Weald was begun, and a great deal of erosion of Chalk took place, giving rise to a surface known as the sub-Eocene surface which can still be seen today (Fig. 6). Accumulation of marine and

estuarine deposits in the early London and Hampshire basins followed, covering a wider area than the present outcrops of Eocene
rocks, which are the remnants of these deposits. Next came a period
of more intense folding and faulting, accentuating further the form
of the London and Hampshire basins and the Weald, and bringing
about a series of minor folds and faults trending east–west across
the Weald, Hampshire, Salisbury Plain and the London basin. Thus
none of these areas has a simple structure, as illustrated in the
sections, Figures 8 and 11.

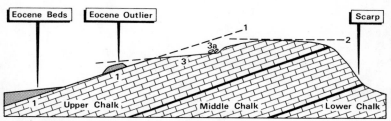

Fig. 6. *Sketch of different facets of a typical chalk scarp. 1–1, the sub-Eocene surface,
the oldest, trimmed before the Eocene beds were deposited; 2, the late Tertiary surface;
3, the early Pleistocene marine bench and cliff; 3a, deposits of the same, e.g. Netley
Heath Deposits.*

This folding and faulting took place at the same time as the main
Alpine folds of Europe and elsewhere; indeed it has been described
as 'the outer ripples of the Alpine storm'. Since then most of Britain
has been nearly stable, subject only to the action of weathering, ice,
rivers and the waves of seas at various levels to carve the scenery.
The main exception to this is the steady subsidence of the North
Sea basin and parts of eastern England. Following the folding,
much of south-east England was reduced to a level surface by rivers
in the passage of millions of years; remnants of this surface now
stand at about 700 to 1,000 feet above present sea-level. These
remnants may be recognized today in the highest parts of the Chalk
and of the Cotswolds (Fig. 6). Where the Chalk has not been overrun by ice then this old land-surface (referred to here as the late
Tertiary surface) has been exposed to the weather for the millions
of years since its formation. During this time deposits known as

Plate 6. *The North Downs east of Guildford. The chalk escarpment here (on the left) has a cover of superficial deposits, and is well wooded, in contrast with the area of the South Downs shown in Plate 5. The distant chalk skyline is Netley Heath, an area planed off by the early Pleistocene sea. The area to the right of the chalk scarp is mostly on the various rocks of the Lower Greensand: here the Upper Greensand and Gault Clay are thin and have narrow outcrops.*

clay-with-flints have been formed. As these deposits yield heavy, brown, wet, acid soils, and carry damp woodland as a result, many of the level summits of the Chilterns and North Downs are heavily wooded and are very different to the ideal picture of grassy downs (Plate 6).

In early Pleistocene times (following the Tertiary era – see Chapter 22) the sea invaded the London and Hampshire basins, and rose to a height of nearly 700 feet above present sea-level. Its greatest extent is now shown in places by a much subdued cliff-like feature at this height, sometimes notching the dip-slope edge of the late Tertiary surface described above (Fig. 6). Occasional beach deposits of this sea have been found, particularly at Netley Heath in Surrey (Fig. 6). Below the old cliff is a slope which marks the former sea-floor and this slope has also been exposed to the weather long enough to carry some damp clay soils and woodlands.

Below this again can be found areas where the former cover of Eocene rocks has been eroded away, exhuming the original sub-Eocene surface, which often carries soils suitable for arable farming, as for instance in east Kent. This surface, if it is found, will pass downwards in turn beneath a complicated series of Tertiary rocks, and gravels and terraces carved by rivers, with which we are not concerned in this chapter. The ideal relationships between the sub-Eocene surface, the late Tertiary surface and the early Pleistocene marine bench are shown in Fig. 6.

The Pleistocene ice sheets have left boulder clay on some of the northern chalk outcrops, and modified and lowered the hill shapes. In Breckland, in Norfolk and Suffolk, for instance, the Chalk is generally below 200 feet above present sea-level and covered with glacial sands and gravels.

The Chalk forms upland country in the scarplands – the highest point on the English Chalk is Walbury Hill in the south-western tip of Berkshire at 975 feet. It is pertinent at this stage to ask why this should be so, for Chalk is a comparatively soft rock. It is mainly because of the porous nature of the Chalk; nearly all the rain that falls on Chalk sinks below ground to contribute to a reservoir of ground water, and so there is little water and river erosion at present.

This ground-water reservoir will best be understood if it is realized that rain water sinking into the ground passes downwards until it is held up by some impervious layer, or until it joins another body of water similarly held up. In the Chalk, which has few impervious layers, there are often great bodies of ground water supported at depth by an impermeable bed; for tens, perhaps hundreds, of feet above this all the rocks are full of water, the pores and crevices being completely occupied by water which is able to percolate through the whole in the direction of any outflow, such as a spring. Above this water-saturated zone there is usually a damp zone, in which the only water is recent rainfall on its way to join the main mass. Between these two zones is a more or less definite plane of demarcation which is known as the 'water-table'; below this the rocks are normally saturated, above it air is present in pores and crevices. The level of the water-table at any place varies according to the supply. During or after wet spells it rises, while dry spells may cause it to sink. Owing to the time taken by water in passing through the minute pores of these rocks, there is frequently a great lag between the time of maximum rainfall and maximum water-table level.

Chalk Valleys

This rise and fall of the water-table has effects on stream flow. A small proportion of the valleys in the Chalk now carry streams or rivers; many more are dry, having no stream at present. A further number come into an intermediate category, being occupied by streams only during or after very wet seasons. Such temporary streams are known as 'bournes' or 'nail-bournes', and the frequency of such place-names as Bourne or Winterbourne on chalk areas illustrates their distribution. These are the valleys normally just above the water-table, but after a spell of heavy rainfall this water-table may rise sufficiently to bring water into a normally dry valley, and the bourne is then 'up'.

The dry valleys pose many interesting problems. Their form and arrangement are so similar to the few stream-bearing valleys that there can be no doubt that they were carved by streams when surface water was more abundant: in many of them deposits of alluvium and gravel swept down by the rivers still cover the floors. They may have been caused by streams flowing when the water-table was higher than at present, but why should the water-table have been higher? It could be due simply to increased rainfall, or to circumstances suggested by Mr C. C. Fagg. He has pointed out that, as the scarp is eroded and recedes, the vale beneath is simultaneously lowered, causing a lowering of the spring line at the junction of the lower impervious beds and the Chalk, and this in its turn will lower the water-table in the Chalk (Fig. 7). Thus, recession of the chalk scarp may well have caused the water-table to be lowered beneath the valleys, leaving them dry.

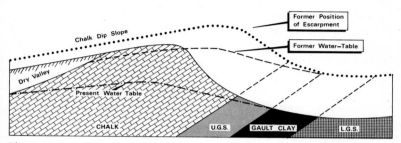

Fig. 7. *Section of a chalk scarp, showing the water-table and a dry valley, and illustrating one possible mechanism for the formation of dry valleys, proposed by C. C. Fagg. U.G.S. is Upper Greensand, L.G.S. Lower Greensand.*

An alternative view is also possible, that the dry valleys were formed when the surface soil was frozen (at least during certain times of the year) during periods of colder climate, especially during the Ice Age. This would result in less percolation of rain water and more activity by surface streams, some of which were probably nourished by snow-melt from the hills. Great spreads of hill-top flint gravels were washed down the valleys where they mixed with finer materials which had sludged down the hill slopes in the freezing conditions. Fans of these deposits, referred to as Coombe Rock, may be seen at many of the dry valley mouths, especially near Wallingford in Berkshire, at the foot of the South Downs (the Black Rock Cliff at Brighton shows this well), and at the foot of the gaps through the Chilterns at Wendover, Tring and Princes Risborough.

Some of these dry valleys have very steep head walls, and frequently such valleys have sharp bends in their courses. Examples are the Devil's Dyke near Brighton, the Coombe at Ivinghoe, and the Pegsdon series of valleys near Hitchin. Mr W. V. Lewis suggested that these features could be accounted for by earlier spring sapping, with a higher water-table giving a spring which eroded the steep valley head with the passage of time. The acute bends in the valleys may be explained by the former streams following the lines of joints and fissures in the Chalk. It is probable that any particular dry valley may have originated as a result of some combination of several of these hypotheses.

The water-table follows in a subdued fashion the shape of the surface above, and thus the ground water is generally nearer the surface in the valleys than on the uplands above. Thus water can be obtained from shallower wells in the valleys, and this helps to account for the preponderance of valley settlements. Dew ponds were formerly constructed on the uplands to supply water for the animals pastured there.

The Rivers of the Chalklands

The courses of the rivers in the chalk areas are also of interest. First we may note the gaps cut right through the escarpments of the North and South Downs by a number of rivers, namely the Wey, Mole, Darent, Medway and Stour through the North Downs, and the Arun, Adur, Ouse and Cuckmere through the South Downs. The rivers obviously began cutting their gaps when the land of the Weald inside the Chalk stood as a complex dome, higher than the Chalk itself. Later erosion of the less resistant beds inside the Chalk scarps, while the gaps through the Chalk were cut down rather than widened, has left the present situation. These rivers will be considered in more detail in the chapter on the Weald.

Gaps through the main western chalk scarp are not so frequent, but some are more spectacular. The major gaps cut by rivers flowing from west to east are the Goring Gap of the Thames, and the Humber gap, while the much wider breach of the Wash has the four Fenland rivers draining into it. This last gap may have been lowered greatly by ice action. Smaller rivers cut through the Dorset Chalk, namely the Stour and Frome. The course of the Salisbury Avon is described below.

There are further gaps in the chalk escarpment which are dry, having no river today. They are well below the scarp level and their form indicates that they are probably former river valleys. These include the Compton gap through the Berkshire Downs a few miles west of the Goring Gap, and the Chilterns gaps at Wendover and Tring which carry a canal, railways and roads through the Chalk. The origin of all these gaps, dry and those carrying rivers, may be sought in the south-easterly drainage pattern which was established following the uplift which marked the end of the chalk sea and deposition, and the initiation of the English scarplands. While some of the gaps still carry streams, the activities of rivers working in the more easily eroded clay vales beneath the scarp has robbed others of their through drainage.

The gaps must have originated, then, when the clay vale stood above the present position of the chalk scarp, and while the narrow gap or gorge through the Chalk is impressive, one should perhaps be more impressed by the amount of material that has been removed from the clay vale.

Other remarkable river courses in the chalklands are those of the Hampshire and Wiltshire rivers which flow in southerly directions across a series of east–west folds and faults. These rivers include the Salisbury Avon, the Test, Itchen and Meon. It is thought that the folds in the area, of which only subdued traces can be seen today, were planed off by the early Pleistocene sea advancing across the area up to a level of about 700 feet above present sea-level. On the retreat of this sea, the rivers began to flow southwards over the emerging smooth sea-floor, and they have maintained these courses since, with minor adjustments to structure.

The Salisbury Avon in fact rises in the Vale of Pewsey, an anticlinal vale cut into the main western chalk escarpment (about which more will be said below), where there is an indeterminate watershed between this river and tributaries of the Bristol Avon. The Salisbury Avon then cuts through the southern scarp of the Vale of Pewsey, following the southern limb of the anticline, and from there continues its southwards course to the sea in Christchurch Harbour.

Flint and Sarsens

A notable feature of Chalk is the presence within the rock of flints. They chiefly occur as isolated nodules a few inches or more across which are generally arranged along the bedding. The flints are black or brown, the outsides being white where they merge into the Chalk. They are composed of silica, an insoluble material far harder than the surrounding Chalk, and consequently the disintegration or solution of the Chalk in which they occur leaves the flint nodules behind. Thus flints originally derived from the Chalk may be met with over wide areas from which much or all of the Chalk has been removed, forming flint gravels in river valleys and flint shingle beaches along many coasts.

The silica comprising the flints has probably been derived in the main from the skeletons of sponges which lived in the seas where the Chalk was deposited. Occasionally a flint encloses a fossil sponge, and preserves its character, but quite frequently flints surround shells or sea-urchins, the 'lime' or calcium carbonate of their skeletons being so replaced by the silica that all the minute details of their structure are retained in the flint. From such evidence it is apparent that the flints are not original deposits of silica formed as such on the sea-floor; they represent 'concretions' of secondary origin, the silica having been disseminated widely in small particles through the whole mass of the Chalk until some time after its deposition. When the Chalk was finally elevated from beneath the sea, the silica was concentrated by solution and precipitation in rhythmically arranged layers more or less following the beds of the Chalk.

Flint has exercised more influence on early human history than almost any other rock in Europe. For early man discovered that by breaking flint he could obtain sharp-edged tools suited to his many simple needs. And for tens of thousands of years, even after he had acquired considerable skill in the making of implements, he continued to use flint, chipping and flaking it to form a wide variety of tools and weapons. For this reason, as for others, the chalk areas attracted great prehistoric populations, and the mining of flint became an important industry, for example at Grime's Graves near Brandon, on the border of Norfolk and Suffolk.

Among the inhabited regions of prehistoric England none was more important than Salisbury Plain. The reasons for this are not far to seek; among the lowlands and vales of southern England the

Chalk ridges were used as natural causeways, especially after the climate became mild at the close of the Ice Age. Then the clay vales became marshy and densely wooded, commonly by oak, becoming practically impassable to primitive man. At the same time, woods on the thin, dry soils of the Chalk would probably be easier to clear and penetrate, and a system of trackways led to Salisbury Plain from several parts of the south and east coasts, along the radial chalk outcrops already described. The plain thus became a focus of routes each following the dry and comparatively open uplands, and it is not surprising that this region shows more signs of early occupation of different dates than almost any other area, with its ancient trackways, its hill-top camps, its tumuli and its stone monuments.

Stonehenge is the best known of these stone monuments, although smaller and more modern than the circle at Avebury. The larger stones of Stonehenge and all those of Avebury were derived from sarsens, stones of local origin (see Glossary) which are also known as grey wethers, Druid stones, or bridestones. These sarsens are found on many parts of the Chalk, but nowhere so common as on the downs near Marlborough, especially in valley bottoms in the neighbourhood of Fyfield Down, where hundreds are to be seen. South of the Kennet they are less abundant, but similar stones are found beyond the limits of the chalk country as far as Swindon. Many of the blocks are of great size, and many are covered with lichen and lie half buried. Among the Avebury stones are sarsens over twenty feet in length probably weighing more than sixty tons, but which must have been transported for miles and arranged to form one of the most imposing prehistoric monuments in Europe, with a picturesque village that grew up much later near the margin of the circle.

The larger stones of Stonehenge are sarsens of local origin, dressed and sized, but others have been brought from the Prescelly Hills of Pembrokeshire. The whole area is obviously an early centre of great importance. Near Avebury stands Silbury Hill, the largest artificial mound in Europe, and the Kennett Long Barrows, the largest in Britain, as well as many other monuments, including Windmill Hill, Neolithic causewayed camp of the earliest farmers of England.

It is remarkable that the sarsens should have yielded masses of durable stone in a state which may be described as ready quarried in this area of Chalk, for the Chalk as a whole is singularly deficient in building materials. It is doubtful whether the early stone circles would have been built on Salisbury Plain without the sarsens, or without the flint implements which were the tools that raised the economy above the meanest subsistence level. Flints also have been used as building material in recent centuries, although the small size of most of the flints and the difficulty of dressing them have led to the development of peculiar and localized architectural styles. In some parts harder bands of Chalk have been used for building, so that cottages of Chalk may be seen, as at Blewbury in Berkshire. Wood is also extensively used, particularly for farm buildings.

The Central Area of Chalk

It now remains, in the light of the foregoing general remarks, to make a very brief regional survey of the various tracts of chalk-land. This must logically begin with the large central area of Wiltshire and Hampshire. This is the great centre of prehistoric England just described, and with its extension into north Hampshire as far as Selborne, it makes an area almost fifty miles from east to west. The Marlborough Downs are separated from Salisbury Plain by the Vale of Pewsey, a low-lying inlet which intrudes into the chalklands eastwards from Devizes. From the northern scarp of the Marlborough Downs overlooking Swindon to south of Salisbury, where the Chalk dips beneath the newer rocks of the Hampshire basin, is just over thirty miles. These wide uplands are sparsely populated areas of large fields and occasional clumps of woods, often beech. There are wide areas of bare Chalk, with deep valleys, particularly in Salisbury Plain, but great diversity may be shown where any superficial cover is present. Thus there are stretches of gravel on Salisbury Plain, tracts of rough pasture, sometimes with gorse. Elsewhere there are woodlands and damp pastures overlying clay-with-flints. Particularly is this true of the summits of the Hampshire Downs near Selborne.

This is a country of minor east–west folds with little surface expression today, with drainage mainly southwards across these folds. The rivers have achieved varying degrees of adjustment to these folds, generally more in Hampshire than in Wiltshire, shown by the relative number of east–west streams.

An examination of the scarp slopes bounding this area is instructive in showing some of the characters of chalk structures. The main northern and western edge is frequently a double scarp. This is caused by differences of hardness within the Chalk itself, which is not so homogeneous as it appears at first sight. Certain horizons are better able to resist erosion than others, so that subsidiary scarps within the chalk areas are sometimes as prominent as those which bound it. A good example of this can be seen north-west of Marlborough, where a very steep scarp at Clyffe Pypard marks the edge of the Chalk. This scarp fronts a bench of Lower Chalk, and behind this is another steep scarp, that of Hackpen Hill, with the slope in Middle Chalk and the crest of the scarp formed by Chalk Rock, a thin, hard band which occurs between the Middle and Upper Chalk. This scarp form is often found, particularly along the front of the Berkshire Downs and the Chilterns, and is illustrated in Fig. 6. In front of this Lower Chalk bench there is in places a further less prominent bench of Upper Greensand, the rock beneath the Chalk, before the clay vale is finally reached. Around Harwell and Didcot the soils of this Greensand bench carry fruit orchards. Along this western scarp, farms and villages have been located at regular intervals near the spring line where the waters of the Chalk or Greensand are thrown out by the impermeable clays beneath (Gault or Jurassic, depending on the detailed succession in a particular area).

Minor folds within the Chalk are well illustrated by the chalk edge where the Hampshire Downs meet the London basin west of Basingstoke. Normally one would expect the Chalk to dip beneath the younger rocks of the basin with no marked change of relief, and this is so between Odiham and Basingstoke, shown in section in Fig. 8. But going farther westwards, one sees a chalk scarp overlooking the London basin, at first thought a completely anomalous feature. This scarp marks the southern limbs of a series of small anticlines or upfolds here, the northern limbs having a steep dip

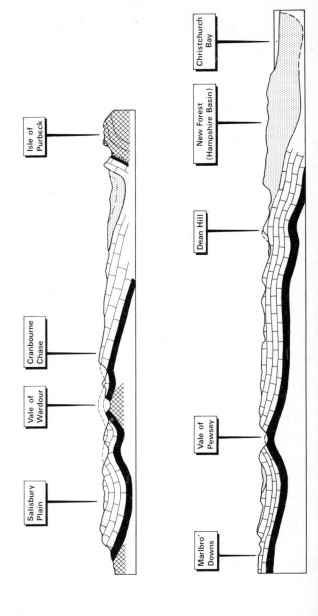

Fig. 8. *North–south sections through southern England, showing the various structures.*

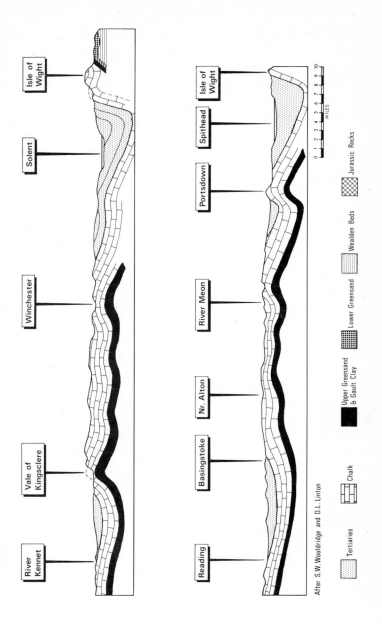

River Kennet — Vale of Kingsclere — Winchester — Solent — Isle of Wight

Reading — Basingstoke — Nr. Alton — River Meon — Portsdown — Spithead — Isle of Wight

After S. W. Wooldridge and D. L. Linton

0 1 2 3 4 5 6 7 8 9 10
MILES

Tertiaries

Chalk

Upper Greensand & Gault Clay

Lower Greensand

Wealden Beds

Jurassic Rocks

and little surface expression. In front of this scarp small patches of Greensand outcrop near Kingsclere and Ham, marking the denuded centres of these anticlines (Fig. 8).

The line of these anticlines is continued westwards in the Vale of Pewsey (Fig. 8). This deep inlet extends eastwards from Devizes for almost twenty miles, carrying into the chalk upland a narrow belt of lower country which is flanked on both north and south by a chalk escarpment. The Vale of Pewsey has a hillocky floor of Upper Greensand, and many villages with large and beautiful houses of brick and half-timber, but it is not a real river valley. It is drained almost entirely by the Salisbury Avon and its tributaries; this river turns south about half-way along the vale and cuts through the chalk scarp at Upavon, to become a southerly flowing stream cutting across small folds as far as Salisbury. The vale has been produced by the action of the tributaries of this and possible earlier rivers along an east–west belt where the Chalk is thrown into a gentle anticline, which has proved weaker than the undisturbed Chalk to north and south (Fig. 8). The axis of the anticline is actually under what is now the northern edge of the vale, so for that reason southward-flowing streams may have been expected. Tributaries developed rapidly and cut through the Chalk, exposing the Greensand beneath, and the Avon succeeded in taking over the drainage of the Vale.

The Vale of Wardour farther south (north of Shaftesbury) is a somewhat similar inlet into the Chalk developed along a comparable weak anticlinal belt, but the removal of the Chalk has here exposed a varied series of rocks including Greensands and Jurassic clays and limestones which, enriched from downwash from the chalk scarp, have given rise to varied soils and to an agricultural area of great beauty (Fig. 8). This vale differs further from the Vale of Pewsey since the river Nadder flows through it from west to east, in accordance with the east–west folding, joining the Avon at Salisbury.

It is noteworthy that both these vales have very indistinct watersheds at their western end. The country west of Devizes and the Vale of Pewsey is drained by tributaries of the Bristol Avon. The Kennet and Avon canal used the Vale of Pewsey to progress from one river to the other; a tributary of the Kennet rises at the extreme eastern end of the vale, and at the western end a steep descent was made to the Bristol Avon drainage by the Seend flight of locks. If we stand at these derelict locks today, it is not difficult to imagine that, given sufficient time, the Bristol Avon tributaries could perhaps cut back headwards and capture the drainage of the Vale of Pewsey from the Salisbury Avon.

The Vale of Wardour has a very poorly defined divide, not marked by any great difference in height, between the drainage of the Nadder and the Stour. Whichever river the water eventually enters, it will reach the sea through Christchurch Harbour, into which both the Stour and Salisbury Avon debouch.

The Dorset Chalk

South of the Vale of Wardour the Chalk outcrops in Cranborne Chase, with its scarp facing west and its dip slope running down to the newer rocks of the Hampshire Basin (Fig. 8). Beyond the river Stour gap at Blandford Forum the scarp continues, but becomes rather disjointed, near the uninhibited Cerne Abbas giant carved in the Chalk, and so to Eggardon Hill. Beyond this point extensive outcrops of Upper Greensand lie in several patches unconformably over various older rocks, and nearly horizontally at about 800 feet above present sea-level, making the level-topped hills around Honiton, in Devon, and rising to 1,000 feet in the Blackdown Hills, north of that town. Some of these patches of Greensand have Chalk upon them. The Chalk and Upper Greensand reach the coast between Lyme Regis and Sidmouth, being prominent in Beer Head. A further isolated outcrop of Upper Greensand at the same height is found in the Haldon Hills south of Exeter, these hills being capped by flinty gravels of Eocene age. The whole of this Chalk and Greensand area has been planed off, first by the sea before the 'Alpine' earth movements, which caused only a gentle flexuring and some faults here, and by sub-aerial agents afterwards.

The Dorset Chalk also extends eastwards in a narrow outcrop from Dorchester to Swanage. At Swanage it is interrupted by

Bournemouth Bay, but appears again in the Isle of Wight. This outcrop is dealt with in the south coast chapter (Chap. 7).

The Hog's Back and the North Downs

This chalk outcrop separates the London basin and the Weald, with younger rocks to the north, and older rocks to the south. The Hog's Back is between Farnham and Guildford, and here the Chalk dips nearly vertically, giving only a narrow outcrop and a well-defined, nearly symmetrical ridge rather than a conventional escarpment. The steep, narrow ridge is accentuated by a fault to the south of the chalk outcrop (Fig. 11). Where nearly vertical Chalk is not faulted, the topographic expression may be very subdued compared with an escarpment of gently dipping rocks. Examples of this may be seen by comparing the northern and southern limbs of the Kingsclere anticline described above. The northern limb dips steeply and is marked by a slight swelling of the ground, while the gently dipping southern limb gives rise to a strong scarp.

East of Guildford, the North Downs become a conventional escarpment, with its scarp facing south over the Weald, and its dip slope towards the London basin and Thames estuary. Minor folds bring Chalk to the surface again near Erith, Purfleet and Tilbury, and also in the Isle of Thanet, in the north-eastern angle of Kent. The latter is not now an island, but was so in historical times (see (p. 101).

The escarpment of the North Downs is cut by river gaps, each of which is the site of at least one town. Guildford on the Wey, Leatherhead and Dorking on the Mole, and Canterbury on the Stour are examples, while several smaller towns are found in the Darent gap, and larger ones on the Medway. The summit of the Downs is generally well wooded, for it carries a cover of clayey superficial deposits. The highest portion of the North Downs lies between the Medway and Mole, and it is thought that here remnants of the late Tertiary surface survive (Fig. 6), while beyond this the early Pleistocene sea covered the present summit. There are stretches, therefore, of this marine bench as well. Below this, particularly in east Kent, the sub-Eocene surface is also widely exposed, and this area supports varied agriculture. These surfaces, particularly the higher ones, are extensively cut into by dry-valley systems, and the ancient routeways such as Watling Street avoided both the woodland and the dry valleys of the crests and kept to the lower slopes.

The eastern end of the North Downs is splendidly marked by the white cliffs of Dover.

The South Downs

At Butser Hill, south-west of Petersfield, the Hampshire and South Downs meet, and the South Downs escarpment turns away eastwards, with the scarp looking north over the older rocks of the Weald. Minor folds in the Chalk account for the chalk hill of Portsdown above Portsmouth, and also for the delightful scenery around Goodwood and the Trundle. From Butser Hill to Arundel the Downs are rather higher and more wooded than they are farther east, although here there is but little superficial cover. Arundel occupies the westernmost of the four river gaps through the South Downs, and here the Chalk gives rise to steep river cliffs. East of the Arun gap the Downs, apart from small high areas, were overrun by the early Pleistocene sea. From the Arun to Beachy Head, the South Downs are largely without superficial cover, and considerably dissected. Here there is much of the country, thought so typical of the English Chalk but rarely found, of 'blunt, bow-headed, whale-back downs', with its short grass cover of springy turf (Plate 5).

The South Downs also end splendidly in chalk cliffs at the Seven Sisters and Beachy Head, the latter being over 500 feet high. Chalk cliffs are usually nearly vertical, presumably because the porous nature of the rock prevents gully erosion on the cliff face, and in places possibly because of joint planes in the rock itself. The Seven Sisters are fine examples of vertical chalk cliffs, deriving the name from the truncated valleys and spurs seen in section along the cliff.

The Western Chalk Scarp

The Dorset scarp, the Vales of Wardour and Pewsey, and the scarp of the Marlborough Downs (near Clyffe Pypard) have already been described. The chalk scarp sweeps majestically on from here in a westerly and north-westerly direction, at first fronting the Berkshire Downs as far as the Goring Gap. These downs are breezy, arable uplands for the most part, along the summit of which runs the green, prehistoric Ridgeway, with a chain of Iron Age hill forts along its line. The highest of these is Uffington Castle, near which is the white horse, carved in the scarp slope, which gives its name to the Vale of White Horse below. These figures carved in the Chalk testify to the thin soil found on the bare white rock. This particular horse is of considerable antiquity, probably having first been cut by a Belgic tribe about 100 B.C., but most of these figures date from the eighteenth and nineteenth centuries, while the one at Pewsey was cut only in 1937. Some are visible from long distances, for example that at Westbury, 'a very modern well set cob', which can be seen from many points on the Mendips.

Beyond the Goring Gap there are the Chiltern Hills, with well-developed slope facets (Fig. 6). A flat, wooded, late Tertiary summit plane, cut into by dry valleys, sits above the scarp formed in the Middle Chalk and a well-developed Lower Chalk bench at its foot in one direction, while the early Pleistocene marine bench and sub-Eocene surface can be seen before the Eocene outcrops of the London basin are reached in the other direction. The dry gaps through the Chilterns have been mentioned earlier. Proceeding north-east, the summit levels become lower and woods are fewer; beyond Luton 600 feet is rarely exceeded. We reach the area which has been overrun by ice, and superficial cover, where it occurs, is of glacial origin and very varied. So we travel on in very subdued chalk country, past the Gog Magog Hills near Cambridge and into East Anglia. The Chalk here is very low; in the Brecklands it rarely exceeds 200 feet in height, and its cover of glacial sands and gravels leads to heath cover and, lately, Forestry Commission plantations. Farther north in Norfolk good agricultural land covers the Chalk.

Across the Wash, the Chalk once more gives rise to an upland area in Lincolnshire, its western escarpment bounding the clay vale of the Ancholme and Witham while its eastern edge is marked by an ancient cliff overlooking the wide coastal tract of marsh and clay. The scenery of this area presents some features comparable with those of the Chilterns and southern areas. In the south the scarp is deeply fretted, but farther north near Claxby and Caistor, although the wolds are not so high, they present a sharper face to the west; in this area, too, there are few surface streams, but such villages as occur are small and are mostly found in the dry valleys. On the uplands, as in most chalk areas, the farms are large, with extensive yards and granaries. Farther south the wolds show a different type of scenery, the Chalk being covered in many places by deposits of boulder clay, sand and gravel: these give rise to conditions of soil and agriculture quite unlike those of the bare Chalk. These differences have been brought about by deposits left by the ice which formerly covered the whole of this tract; since the ice melted, the material it had deposited has been removed from some parts by rivers, but sufficient time has not elapsed since the Ice Age to allow all these deposits, which though they vary in thickness are only superficial, to be carried away.

The Chalk is widely breached by the great gap of the Humber, but the Yorkshire Wolds represent its further continuation. The wolds, however, differ greatly from other chalk areas in England, for the slopes are more abrupt and the lower parts are more widely cultivated than in the downlands. These characters result chiefly from the fact that the Chalk of Yorkshire is harder than that of the south of England, many beds being more like ordinary limestones: thus they stand out more as impressive cliffs and lack the rolling hills of the south Chalk country.

The Chalk of the Yorkshire Wolds may be said to end in Flamborough Head, a fine headland now rather spoilt by untidy development. This is the northernmost chalkland in England.

Holderness

The land south of Flamborough Head to the Humber was a wide
bay before the Ice Age, but during the glaciation a thick spread of
boulder clay and other glacial deposits filled this bay and these
deposits now form the stretch of country known as Holderness.
The old cliff marking the edge of this former bay can still be traced
at a height of about 100 feet along the edge of the wolds from
Sewerby, near Flamborough, to Hessle on the Humber. The cliff
can also be traced across the Humber in Lincolnshire, where
country similar to Holderness fronts the Lincolnshire Wolds, passing to the south into the Fens.

It may seem inappropriate that a tract of low boulder-clay
country should be described in a chapter on the chalklands, but
Chalk underlies the boulder clay, and Holderness can logically be
included nowhere else in the division scheme here adopted.

Holderness is now a tract of boulder-clay country about twelve
miles wide, with an irregular surface. Occasional tracts of marshland and mere were present, and a mere a mile and a half long still
remains at Hornsea, while the sites of others are recognizable at
Bridlington, Skipsea and Withernsea, or are indicated by such
place-names as Marfleet and Rowmere.

Along much of the coast the low cliffs of red or purple boulder
clay are rarely more than thirty feet high. As they are easily eroded
by the sea, they are receding at an overall rate of some two or three
yards each year, and several villages have been lost in historical
times. Though this land was only added to England as recently as
the Ice Age, no part of the British coast is being destroyed more
rapidly.

But great as the losses have been along the seaward cliffs, there
have also been gains elsewhere. Undoubtedly some Holderness
material must contribute to the periodic growth of Spurn Head,
while inside the Humber estuary man and nature are combining to
bring about some reclamation. The village of Hedon, which formerly was a flourishing port, is now two miles from the coast,
with its square church tower standing conspicuously above the flat

lands, and its old docks and water-ways traceable among the green fields.

The Holderness lowland has only one river of its own, the Hull, at the mouth of which stands Kingston-on-Hull, the isolated city of the Humber. As in many parts of the more southern fenlands, bricks are much used in building, and most villages have thatched, brick cottages, while bricks occasionally have found their way into church-building, as in Holy Trinity, Hull. Along the coast, however, many old churches and houses have been built from boulders derived from the boulder clay, particularly when they could be gathered from the beaches, and they then show much variety in colour.

5. The London Basin

An expanse of low ground extends about sixty miles west of London to Newbury, drained by the Thames and its tributaries, particularly the Kennet which rises on the Chalk near Avebury and flows east to join the Thames at Reading. This low ground is bounded to north and south by chalk uplands, to the north the Berkshire Downs and the Chilterns, and to the south the Hampshire Downs, the Hog's Back and the North Downs. Travelling west from London, the Chalk gradually closes in, pinching out the low ground of the basin. Within this area of low ground many types of scenery are represented. The wide green meadows where 'the silent river glides by flowering banks' (Plate 8), the low marshes along the estuary, rich agricultural lands, and dry heaths are all in contrast with the surrounding uplands of Chalk. The centre of the basin is occupied by the great expanse of the city and its suburbs, spreading outwards along all the main routes. The description of these miles of brick and stone is no part of the present purpose, but we shall presently note in passing how its foundation and much of its later growth were conditioned by geological and physical factors. Its building materials have been drawn from a wide area, but much of London is built of dull yellow bricks which supply the dominant tint to many parts: many of these bricks were made from the London Clay which occupies more of the basin than any other rock (Fig. 9).

In Essex this clay also makes up large areas, and here brick has long been the dominant building material, this county being particularly noteworthy for the use of brick in church-building from an early date. For bricks were not frequently used in churches in other areas until the last few centuries, even in those areas which are comparatively stoneless. These clay tracts were formerly heavily wooded, and wood has been used in Essex churches more than elsewhere, the wooden spires (for instance at Blackmore) being very characteristic.

In modern times the construction of buildings and roads from reinforced concrete has meant that thousands of acres of the Thames valley have been stripped of their gravels, leaving on the flood plain artificial and frequently unsightly lakes, although occasionally these pits are left in a creditable state.

The explanation of the variety of scenery must be sought, as elsewhere, largely in the past. The London basin is a structural basin (Fig. 5). After the Chalk was laid down as a more or less level sheet, erosion and warping took place and a trough was formed. Water filled this trough and sediments were laid down over a much wider area than their present extent. Earth movements continued during the era known to geologists as the Tertiary, which followed Chalk times. The basin was filled with some river-borne sediment from the west, and marine or estuarine material from the east. As earth movements continued and sea-level changed, marine and river deposits were laid down, sandwiched between each other. That part of these deposits which has survived subsequent erosion now floors the triangular London basin between the chalk hills from Newbury to the Thames estuary. Scattered patches of Tertiary deposits survive on the Chalk beyond this main boundary, testifying to their former greater extent. Glaciation has also left its mark on Essex, so that the Tertiary rocks here are covered by sheets of boulder-clay and till, and the London Clay country passes without any very marked change of appearance into the boulder-clay country of Essex and East Anglia. The Thames and its tributaries, and their predecessors, have carved terraces and laid down gravels which have important effects on settlement and scenery. It is to these past influences that we owe much of the variety of scenery within the basin.

The rocks of the London Basin

The London basin, as has already been said, is bounded to north and south by chalk hills. The northern limit of the basin is the comparatively simple scarp of the Berkshire Downs and the Chilterns, separated by the Goring Gap through which the River Thames enters the basin from the north. These hills dip to the south and south-east, and together may be regarded as one of the succession of north-west facing scarps of the English scarplands, described in preceding chapters. On the other hand, the Chalk of the North Downs dips to the north, and the Chalk extends under the basin, as has been proved by many deep wells sunk in the London district (Fig. 5). The North Downs generally dip more steeply than the Chilterns; the basin is asymmetrical, with its axis generally nearer the southern edge. This asymmetry is further emphasized by the Hog's Back, the westward continuation of the North Downs, with its nearly vertical dip (Fig. 11).

The features of the chalk margins have been briefly described in the previous chalklands chapter, and this will not be repeated here. It is sufficient to refer the reader to those pages, and remind him of the topographical effect of these features and of the varying chalk dip. Exit north and south from the London basin by road and rail is either across boulder clay into East Anglia, or mainly by gaps through the chalk escarpments, with their scarps facing outwards. Minor roads will show the true character of the climb up the chalk dip slope to the summit of the escarpment, and the slope facets of the sub-Eocene surface, the early Pleistocene marine bench, and the late Tertiary surface, with their characteristic soils, vegetation, and topography, which are well developed on the Chilterns and the North Downs, and to a lesser extent on the Berkshire and Hampshire Downs. The deciphering of the complex story of the Tertiary and later history of the London basin was primarily the work of the late Professor Wooldridge.

In places this picture is more complicated, especially on the southern margin. The Hog's Back is a well-defined and nearly symmetrical ridge topographically. To the west, one may travel from, say, Reading to Basingstoke and Winchester, across the basin and the Chalk, without seeing a steep slope, such is the gentle dip northwards of the Chalk; the difference between the two areas is obvious only in the land use. West of this again, an anomalous chalk scarp faces northwards over the basin, due to the minor anticlines described in the previous chapter (Fig. 8).

Another important feature of the Chalk is the minor folds within the basin itself which bring the Chalk to the surface, providing sites which were used for ancient settlement. One such is Windsor, with its castle, while other sites are found east of London – Erith, Purfleet and Grays being among these. The Isle of Thanet is another example which, although far removed, in one sense is geologically part of the London basin.

The Tertiary rocks which lie on top of the Chalk in the centre of the basin consist mainly of clays and sands. The London Clay is the most widespread and uniform of these, with mainly sandy beds lying above and below this clay. As all these rocks were folded with the syncline, those sands below the London Clay occur as a narrow belt adjoining the chalk outcrop on both sides of the basin, while the upper sands form a more extensive spread near the middle of the basin (Figs. 5 and 9).

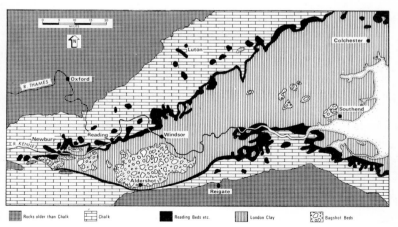

Fig. 9. *The rocks of the London basin.*

The London Clay occupies wide areas of the low country in the basin, stretching from Regent's Park and Wembley northwards into south Hertfordshire, while it also underlies much of the area to the south of the river, reaching almost to Croydon. There are also extensive areas further upstream, for example the Loddon valley south of Reading. The London Clay gives rise to low-lying country with heavy soil. It was formerly well wooded, oaks and elms being the chief trees. In Essex parts of Epping Forest, formerly a royal hunting ground, still survive on it, but in Middlesex most of the woodland has been cleared, though St John's Wood, Enfield Chase and Wormwood Scrubs all testify to the former extent of the woodland. As the trees were cleared the London Clay became used increasingly for pastureland, but on the clay itself settlements were few except where overlying gravels and sands provided drier sites with supplies of water from springs and wells. Even in the neighbourhood of London it was long before dwellings spread over to the clay tracts, which were mostly uninhabited until the beginning of the nineteenth century: modern drainage has made many areas habitable but they are rarely selected for better types of houses, and it is only along the main railway lines that they are completely built over.

The lower sandy beds (Reading Beds and Thanet Sands) occurring along the chalk borders of the basin are often associated with clays. This is because of the oscillation from marine to estuarine and river conditions described above, and thus the character of a particular bed may change from place to place. The sandy beds give rise to various soils of considerable fertility, particularly in Kent where there are extensive orchards and market gardens. Along this belt, too, settlements were above the marshy London Clay country and had readily accessible water supplies, in contrast to the chalk slopes immediately above them: for these and other reasons a line of important towns approximately coincides with it, from Canterbury to Croydon and Guildford.

The rocks of the rest of the basin are well illustrated by travelling south from Windsor. The Great Park, with its woodlands and meadows stretches away southwards on the London Clay, but as the ground rises the scenery changes to wide heaths, often uncultivated for great distances, covered with bracken and heather, or in places with pine woods (Plate 7). Here the gravelly Bagshot sands occupy the middle of the basin, followed by other sandy and loamy beds. These sands extend as far as Aldershot, and their sterile soils provide the Army with some of their training grounds.

Nearer London only small detached patches of these sands remain as outliers standing slightly up above the plain of London Clay, the last remnants of a bed which must formerly have been

Plate 7. *Caesar's Camp, near Aldershot, Hampshire. Heathland on the Eocene sands and gravels of the London basin, much of which in this area is used for military training, the soils being too acid for profitable farming. Here in fact the lower land in the foreground is of clayey Bracklesham Beds, while the steeper slopes above are of Barton Sand.*

continuous with that near Aldershot. Many of the favoured drier uplands are based on these outliers; Hampstead Heath and Highgate reach no great elevation yet stand sufficiently high to afford wide views over the clay plain. Further patches of Bagshot sands are found in Essex, for example the Rayleigh Hills.

The Rivers and Their Deposits

The main rivers draining the London basin are the Kennet above Reading and the Thames below that town. The basin is asymmetrical, with its geological axis nearer the southern limit than the north, but the rivers rarely follow this axis. The Kennet is north of this line, along its whole length, and so is much of the Thames, particularly the great Henley–Marlow loop into the Chalk of the Chilterns. Downstream of London, however, between Erith and Gravesend, the Thames is cutting into Chalk on the southern limb of the basin.

The Kennet–Thames has, nevertheless, an obvious general relation to the basin, and in this it is unique among the larger English rivers which mostly follow the outcrop of some soft bed (frequently the Keuper marls) and occupy valleys which cannot in any way be regarded as structural basins. On either side of the valleys of the Trent and the Severn, for example, the hills are different in structure and rock type, whereas in the London basin a general symmetry of rock distribution and scenic character is determined by the folding.

In detail, however, there are major deviations from this simple conception, some of which have been noted briefly above. Perhaps the most striking anomaly is the Goring Gap, referred to in the to enter the London basin. Above the gap the Thames is a strike previous chapter (p. 60). Here the Thames cuts through the Chalk stream draining the Oxford Clay vale, following the outcrop of this soft bed, and receiving dip-slope tributaries from the Cotswolds. First it cuts through the broken Corallian escarpment at Oxford, and then through the Chalk at Goring. From here downstream the Thames receives the Kennet, and dip streams from the Chilterns, together with tributaries from the south which drain the North

Plate 8. *The Thames and Windsor Castle. The river is here flowing in a flood plain of its own deposits, on London Clay. Local folding brings up a small inlier of Lower Chalk, on which the castle is built.*

Downs and parts of the Weald. Below Goring the Thames remains on the Chalk for about thirty miles. In the so-called Henley loop higher and drier banks afford sites for riverside towns, as at Henley itself and Marlow. Near Maidenhead the river emerges on to the Tertiaries, but soon it again encounters Chalk at Windsor, where a subsidiary anticline or dome brings that rock to the surface as an inlier among the Tertiaries. The Chalk forms a prominent feature on which Windsor Castle stands, and the river has cut into it to form a low but sharp cliff (Plate 8). After this the Thames swings south to Chertsey, and downstream of London it is again cutting through Chalk, here along the southern margin of the basin (Fig. 9).

In considering this seemingly erratic course of the river, we must remember that the present valley is a heritage from times when it was filled with rock, when there was land many tens or hundreds of feet above where we now stand. The Thames is at present expending its energy in widening its valley, meandering freely in flat areas, with a very gentle gradient of a few inches per mile. Downstream it is depositing great areas of alluvium, the tract near Woolwich and Barking being nearly three miles wide.

Occasionally in its meanders the river comes near to the solid rocks on either side of the flood plain, and early settlements frequently grew up at such points, notably Gravesend and Greenwich. In many places the flood plain was formerly a marshy waste, and in the Lea valley there are still considerable areas of marsh, but much of it has been drained to form meadowland. Embankments along the shores in the tidal portion of the river have made this utilization possible, but great areas such as the Isle of Dogs remained as marsh until they were used for dock construction in the last century. In some places east of London the surface of the drained marshlands is several feet below the level of the highest tides, the surface level having fallen, according to the late W. Whitaker, owing to the shrinkage of the peat and clay which compose them. In early times, many of these marshes must have been impassable, and their occurrence on one side of the river or the other for so great a distance from the mouth made it necessary to go upstream as far as the site of London, before a suitable place for a ford or bridge could be found : thus while the solid rocks are near the river at Gravesend, marshes extend on the opposite side, and when at Grays the north bank could be approached, the Swanscombe Marshes made the south bank unsuitable for a crossing.

In Roman times, when the land stood higher relative to the sea than at present, the site of London Bridge was at the tidal limit. Here two small gravel-capped hills stood near the north bank, dry and yet accessible by boat. Here, where the city of London now stands, was the site of Roman *Londinium*. Nearby were other gravel 'islands' rising above the marshes, and their names, such as Chelsea, Putney and Battersea, still have the Saxon ending 'ey' or 'ea' meaning island, although Thorney is now called Westminster. Thus the original site of London was governed by physical conditions.

These gravel 'islands' are fragments of a river terrace, representing the flood plain of a past time when the river flowed at that higher level. Going away north or south from the Thames in London, these terraces are illustrated clearly as the ground tends to rise in a series of small steps, rather than gradually, each step being relatively level (Fig. 10). Thus from the Thames side at Westminster there is a rise to Hyde Park, where the ground is level over

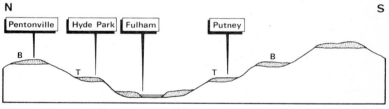

Fig. 10. *The terrace levels of the London area. The terrace gravels are dotted. T is the Taplow terrace, B the Boyn Hill terrace.*

wide areas, though farther north there are other well-marked steps to higher levels. These steps represent the river terraces. The lowest terrace stands about fifty feet above river-level, and thus marks a stage when the land stood some fifty feet lower in reference to the sea than it is now. At this stage the river excavated for itself a wide valley and spread its flood-plain deposits over a great area. These deposits, cut through by the Thames in more recent times when a raising of the land gave the river power to cut vertically once more, have been removed along the area near the river but form a distinct terrace rising sharply from the flood plain; they give rise to the ledge on which Hyde Park stands, an area where the ground is better drained than in the flood plain or in the clay areas. This terrace extends as a wide, flat area from Paddington to Holborn, where it is cut by the deep trench of the Turnmill brook or the Fleet river; thence from St Paul's it spreads eastwards through Bethnal Green to the banks of the Lea. This terrace is known as the Taplow terrace. This and the next higher terrace, the Boyn Hill, are named after districts near Maidenhead, where they are typically developed. Each of these terraces has been subdivided by later research workers and new names have been introduced, but for the purposes of this book only the original names of the terraces most clearly seen will be used.

About fifty feet above the Taplow terrace the Boyn Hill terrace represents a still earlier base-level of the Thames when an even wider spread of gravel was laid down at a level corresponding to the land being about 100 feet lower than it is now. Since its formation much of this higher terrace has been destroyed (during the formation of the Taplow terrace, and subsequently), so that it is mostly represented by fragments of the original level tract, often situated some miles from the river, and in some cases merely forming cappings of small hills. Near London itself, Clapham Common and Pentonville owe their elevation and dry situation to their position on this terrace, which raises them above the London Clay underlying both areas.

Both these terraces indicate a relative elevation of the land since the Thames valley has had substantially its present form. However, there is also evidence of a not very distant period of submergence, for the river alluvium under the present flood plain extends down beneath present sea-level, filling a channel which can only have been cut when the sea-level stood lower than at present. This deeper occurrence of the alluvium has of course no effect on the surface scenery, but the subsidence of the land which has caused it has produced other features in the area of the London basin and elsewhere; the drowning of the valley mouths and the formation of narrow inlets on the Essex shore may be mentioned. These changes in sea-level occurred during the Ice Age; with each major advance of the ice sheets, great quantities of water were frozen into continental ice sheets. rather than existing as liquid in the oceans. and so the sea-level fell. Conversely. when the ice melted. the sea-level rose again in general.

Above the Boyn Hill terrace and farther upstream, evidence is found of another higher river level. This is the Winter Hill terrace of the Thames which corresponds with extensive gravel spreads in the Kennet valley near Roman Silchester. In the Thames valley the Winter Hill terrace is represented by a shallow trench which is cut along the chalk dip slope from Caversham, north of Reading, to Henley, at a height of some 270 feet, and by Winter Hill itself above Marlow. Above this terrace again are higher gravel spreads which run laterally along the Chiltern dip slope. The river and its valley stood at these high levels during the Ice Age, and it has been shown that the Thames flowed at one time along the line of the present vale of St Albans and then across Essex. When this exit was blocked by ice, the drainage was diverted southwards and through the Finchley depression. A further glacial diversion established the Thames in its present course through London.

The ice which caused these diversions also deposited the wide expanses of boulder clay in the north-east of the basin across the chalk dip slopes and the Tertiaries. The boulder clay itself contains much Chalk; for this reason it is a much lighter and more fertile clay than is frequently found in glacial deposits, and where it rests on Chalk the normal characters of that rock are completely transformed by the blanket, which supports some excellent arable land.

A further feature of the eastern half of the London basin is the number of hills which reach about 400 feet above sea-level on a variety of rocks. This is marked particularly around Oxhey, Elstree and Barnet on Eocene rocks, by a platform on the chalk dip slope near Welwyn, on hills near Wargrave, on Hampstead Heath and on the Laindon Hills in Essex on Bagshot beds. Another less well-marked series of hills and platforms occurs at about 200 feet. These 'surfaces' mark the remains of former extensive flats of a higher river Thames.

The London basin is therefore by no means a simple area of low

land. Differences in soil and in relief have been brought about by
the various rock types deposited in the Tertiary era, while the
gently terraced topography reflects stages in the erosion of the
basin. Lacking any striking elevation, such small hills as are left by
the dissection of the more extensive platforms and terraces have
exercised great influence on the settlements of the area, and largely
control those modifications of the scenery which are associated
with the making of London.

The Coast of the Thames Estuary

This is generally a low, muddy and uninspiring coast. Such low
cliffs as occur are generally cut in London Clay and are undergoing
marked erosion in places, as for example the Isle of Sheppey. Much
of the rest of the coast is marshy : salt marsh when exposed to the
sea, and fresh marsh where protected by a system of walls and
banks. Frequently the fresh marsh is lower than the salt marsh outside the banks, for the estuary is undergoing subsidence. This low
land is threaded by the channels of rivers and estuaries which have
been drowned in this subsidence, making a series of islands such as
Foulness and Canvey Island in Essex, and Sheppey and the Isle of
Grain in Kent. The largest of these estuaries, apart from the Thames
itself, is probably that of the Blackwater; this may have been a
former Thames estuary when the land stood much higher than at
present, before the last glacial diversion. It has been mentioned
earlier that previously unusable alluvium areas such as the Isle of
Dogs were used for the London docks in the last century. This
lower Thames marshland here described is now being used for new
major industrial development, such as oil refineries.

6. The Weald of Kent and Sussex

The densely forested area to which the name Weald was originally given still has much woodland in many parts, but most of it is now used for agriculture and as a London dormitory. Its hop fields and orchards, wide meadows and breezy commons, are almost completely surrounded by the chalk downland which forms part of the same structural unit, and in this description, therefore, the North and South Downs may be conveniently included. Crossing this area, dip slope is succeeded by scarp, which in turn is followed by another ridge running parallel to the first. While most of the vales are floored by heavy clays in which the rivers have made wide spreading valleys, the ridges vary greatly in character and their outlines show striking contrasts to the Downs themselves. With such diversity of rock foundation, there are not only differences in soil and agricultural development, but also many types of building materials, and the simple thatched cottages of brick and timber found in some areas give place in others to stone buildings. Over the whole area lies the shadow of the expanding metropolis, so that its character has already been changed completely in the more accessible parts, and its individuality is threatened in many others. Yet even the popularized beauty spots of London's playground retain many features of interest which not even development can obliterate, while there are still considerable areas whose natural loveliness remains. But here, as in much of south-eastern England, it is necessary to get away from main roads to see any unspoiled country.

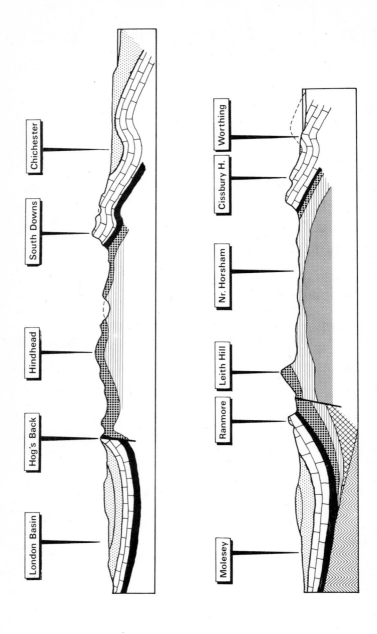

London Basin — Hog's Back — Hindhead — South Downs — Chichester

Molesey — Ranmore — Leith Hill — Nr. Horsham — Cissbury H. — Worthing

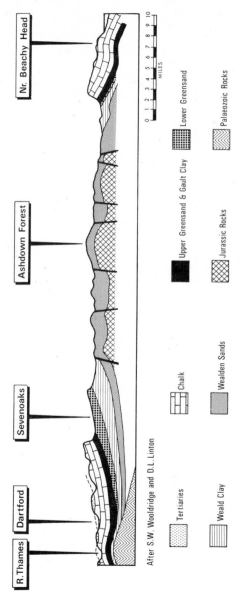

R. Thames | Dartford | Sevenoaks | Ashdown Forest | Nr. Beachy Head

After S.W. Wooldridge and D.L. Linton

Tertiaries

Weald Clay

Chalk

Wealden Sands

Upper Greensand & Gault Clay

Lower Greensand

Jurassic Rocks

Palaeozoic Rocks

MILES
0 1 2 3 4 5 6 7 8 9 10

Fig. 11. *North-south sections through the Weald, showing the various structures.*

The Symmetry of the Weald

The Weald of Kent and Sussex presents a symmetry of structure and a variety of scenery which are in some ways comparable with those of the London basin, for it is traversed by hill ridges running east and west, those of the north of the area matching those of the south. Nevertheless the structure is really the opposite of that of the London basin, for whereas in that area the scarps face outwards, in the Wealden area they face inwards to the centre; the dips are therefore outwards and the Weald is a worn-down dome (Fig. 5). This simple picture should be qualified by saying that the Weald includes many small east–west folds and faults, just as have been seen in the chalklands of Hampshire and Wiltshire to the west, and to a lesser extent in the London basin (Fig. 11).

The view southwards from almost any point on the edge of the North Downs shows the general character of much of the area. For instance from Colley Hill, north of Reigate, the chalk scarp falls abruptly into a vale formed by Gault and Upper Greensand, mostly occupied by permanent pasture but with many oak trees in the hedgerows. Beyond this the land rises slightly to form another scarp, corresponding to the outcrop of the Lower Greensand, an area of arable land, but with much heath and parkland, which extends south-westwards to the more imposing heights of Leith Hill. South of this Greensand ridge another clay plain, more monotonous than the last, stretches for many miles, but out of it the land rises again in another dip slope, to culminate in the High Weald, the core of the region. And far beyond, forming the distant skyline, the clear curves of the South Downs complete the pattern of the area.

The whole structure has been eroded and covered by the sea to the south-east, but may be closely compared with the region across the Channel in France.

Plate 9. *The Weald, a view looking east from about one mile north-east of Hindhead, near the A3. The middle of the photograph shows the low-lying Weald Clay country, while to the left and in the foreground is the Hythe beds escarpment of the Lower Greensand. This escarpment continues to Leith Hill in the far distance.*

The Rocks of the Weald

The rocks of the middle of the Weald are older than the Chalk, but they are newer than the oolites and the clays of the Oxford and Kimmeridge groups. It may occur to the reader to wonder why these beds, occupying so wide an area in the Weald, have received so little attention in the description of other regions. Some of the beds have been mentioned, for example the Greensands occurring under the chalk scarp (p. 54); and in the Vale of Pewsey, which in a small way repeats the structure of the Wealden area, these Greensands form much of the floor (Fig. 8). But in Kent and Sussex a much bigger variety of rocks is present, especially underneath the

Greensands, and these rocks are not much represented in areas farther north in England, for they consist of sands and clays laid down in a lake which extended only over the south-east, a lake of fresh water in which non-marine shell-fish were living.

The sequence of events to be visualized begins with the marine conditions under which the Lias, the oolites and the succeeding clays were deposited, followed by a shallowing of the sea and eventually its disappearance from England; 'Wealden' deposits of sand and clay were then accumulated in a restricted area of fresh water until, with the deposition of the Greensands, the sea once more re-entered the area, the land sinking still farther to allow the sea to cover the greater part of England and to extend, when the Chalk was deposited, over areas which had long been land. The rocks included in the Wealden area may thus be tabulated:

Chalk	
Gault Clay and Upper Greensand	Marine
Lower Greensand	

Weald Clay	
Wealden or Hastings Sands	Laid down in freshwater lake
(including the Wadhurst Clay)	

The Wealden Sands or Hastings Beds, as geologists usually call them, occupy the centre of the area and the newer rocks form more or less regular bands around it, the harder Lower Greensand and Chalk standing out as ridges while the Weald Clay and the Gault Clay form low ground: variations in the thickness of these formations and in the nature of some of the rocks from place to place lead to differences in the heights of the ridges and the breadth of the valleys, and interfere with that complete regularity of scenic pattern which might otherwise be expected.

Although the sea may have washed over much of this region for some time since its denudation began, its present form has mainly resulted from the work of the rivers, aided by frost and rain. These agencies have been able to cut down the outcrops of the softer rocks, leaving the ridges upstanding; these have gradually been undercut and their scarps have slowly retreated to north and south. The Hastings Beds create the High Weald, the area stretching from Hastings to Horsham, a broken upland rising in places to 800 feet above sea-level. This region, sometimes called the Forest Ridges, is mainly an area of sandy soil, once densely wooded. Although the lower slopes are generally of poor grassland, in the valleys there are richer soils. Most of the trees were cut down either for ship-building or in connection with the local iron industry, the ironstone beds in this area being similar to those occurring in the Lias (see p. 37); for many centuries charcoal was used for smelting the ore. Ashdown Forest is situated on this tract, an upland area of woods and commons, but over much of the area only the place-names and the scattered clumps of trees suggest the former extent of the woodlands, while other names like Furnace Pond and Hammer Mill recall the iron industry which helped in the deforesting of the area. In the sixteenth century there were thirty-two furnaces and thirty-eight forges in Sussex alone, but from this area the iron workers moved to other parts of the country, including South Wales, where their activities also led to great tracts of woodland being cut down.

The open commons along the belt through Tunbridge Wells are situated on these sands; in this area the rivers have cut deep valleys. The strong sandstones have been deeply weathered along the prominent vertical joints, and curiously shaped masses stand out from the hillsides, as for example along the valley at Rocks Wood, near West Hoathly; here one mass, 'Great-upon-Little', has been almost completely undermined along a thin bed of clay, a layer only about a quarter of an inch thick but sufficient to hold up the water percolating through the sandstone. Other masses in the same district have been quite undercut and have tumbled over. Similar irregularity is noticeable at High Rocks, a fault scarp near Tunbridge Wells, and on Rusthall Common, where the Toad Rock is a relic of a block isolated by weathering along joint planes and shaped by wind action.

The Weald Clay vales, or the Low Weald, including the Vale of Kent and the Vale of Sussex, almost completely surround the High Weald (Fig. 12). The clays, like those of the Lias and of Oxford, give rise to heavy, sodden soils, and the landscape is featureless. Since the land has been drained or cleared of the oak woods which

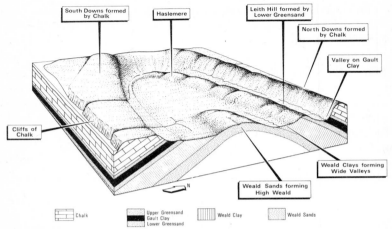

Fig. 12. *Block diagram of the western end of the Weald.*

formerly covered it, it has been converted into neatly hedged cattle pastures or orchards. Trees are still abundant in many parts. Formerly the roads in this clay country were exceedingly bad, and near the end of the eighteenth century oxen were used to draw coaches in some parts because no horse could move in the stiff clay. In this area much clay has been used in making red bricks and tiles, and timber has also been used in building. The major settlements of the Weald are found beyond this unfavourable clay outcrop.

Where the clay outcrops reach the coast they are covered by great tracts of marshy land, much of Romney Marsh marking the seaward part of the Vale of Kent, and Pevensey Levels being similarly placed in the Vale of Sussex.

The Lower Greensand is thicker and more important along the north and west of the area than in the south, and the northern margin of the Vale of Kent is marked by a distinct and steep escarpment, which rises above Ashford and Godalming and culminates in Leith Hill at nearly 1,000 feet above sea-level. This is an impressive ridge, its northern slope gentle and almost the same as the dip of the rocks, its southern face a real scarp with concave, land-slipped slopes rising sharply from the wide, clay plain.

The sandy beds give rise to well-drained upland country with abundant water supplies, and though the soil is sometimes poor, extensive tracts being given over to heathland, there are some fertile belts. Westwards around Haslemere and Hindhead these beds give rise to wider upland areas, with much heathland and pine wood, but the valleys are often more fertile, though the slopes are often steep.

The Devil's Punchbowl, Hindhead, is one of the best-known valleys in Surrey. It has been eroded by a stream flowing north-wards which has cut a valley through the sandy rocks, and the springs along the junction of these with the clay beneath (the Atherfield Clay, part of the Weald Clay) have given rise to the steep slopes of the valley. By the undermining action of the springs the valley head has tended to become larger and wider than the re-mainder of the valley, producing the 'Punchbowl'. In the Leith Hill district are similar steep-sided valleys in which it is easy to trace the spring line and to note the difference in the character of the vegeta-tion where the sands join the clays.

The local variations in this Lower Greensand group lead to dif-ferences in scenery, soils and land use along the outcrop. In the Leith Hill area the Greensand contains exceedingly hard beds of a highly siliceous rock known as chert, a rock very similar in its char-acter to flint. This very resistant material is responsible for the unusual height of these hills. Near Guildford these cherts are less in evidence, and the crest of St Martha's Hill is made up of rather soft sands with masses of ferruginous sandstone (the Carstone), the presence of which has controlled the erosion of these beds, and has here retarded denudation.

Over this Greensand country the picturesque tracks are often deeply cut, in places to a depth of twelve feet or more below the adjoining fields. This sinking of the tracks is still going on, for where the vegetation is destroyed by making a new track over the hill, wind and rain quickly remove the surface soils, and wind action still further deepens the channel as dry debris is produced by the continued disintegration of the rocks at the surface. The Lower Greensand includes many stones which have been used in building, the Kentish Rag in particular being important in the eastern part of the Weald, though it has also been used over a large area, especially for windows and doorways. But almost everywhere along the outcrop of the Greensands from Maidstone to Hindhead some stone houses are to be seen, while the tall towers of the rag-stone churches are also characteristic. The ragstone rocks of the Lower Greensand in Kent break down into fertile soils, used for fruit orchards in contrast with the sterile heaths found on other Greensand areas.

As the ridge dips northwards into the Vale of Holmesdale the Greensand is succeeded by Gault Clay, a stiff blue marine mud, and the scenery again changes. Much of the soil is too heavy for cultiva-tion, save where the admixture of down-washed sand from the Greensand outcrops has produced a lighter soil; elsewhere there are extensive, green pastures. On the south of the Weald the out-crop of the Gault Clay is so narrow that it gives rise to no very important feature (Fig. 12). Above the Gault Clay the Upper Greensand is usually so thin as to be unimportant in relation to our present discussion, occurring merely as a narrow belt on the scarp formed by the Chalk.

The Downs

The Weald proper is ringed by Chalk, except at the coast. The generally inward-facing scarps proceed around the Weald from the white cliffs of Dover via the North Downs, Hog's Back, Hampshire Downs and South Downs to Beachy Head. The main features of these hills have already been described in the chalklands chapter, to which the reader is referred. This section will merely describe some features of particular relevance to the Weald.

The Chalk is here even less used in building than in most areas in England, but flint has been fairly extensively used in Kent from early times, as at Margate and Sandwich.

The North Downs have a steep inward (southward) facing scarp, overlooking the clay Vale of Holmesdale. The scarp is cut by several important river gaps. Its summits are well wooded, due to an extensive cover of clay-with-flints and similar superficial deposits which has accumulated since the summits were planed off either as part of the late Tertiary surface or the early Pleistocene marine planation (Plate 6). The highest parts of the North Downs are over 800 feet above sea-level, while the Hog's Back to the west only just exceeds 500 feet. The western limit of the Weald is the rather irregular chalk scarp in the area around Selborne immortalized by Gilbert White. Beyond this is the continuous plateau-like chalk surface around Alton; no simple dip slope here, the Chalk continues west with its varied scenery to Salisbury Plain. The South Downs, particularly beyond Arundel, are almost free of superficial cover, and these areas of bare Chalk with arable land, springy turf and summit coppices present the idealized chalk landscape so rarely found in fact (Plate 5). The South Downs are likewise cut by river gaps, and we must now consider the details of the Weald drainage.

The Rivers of the Weald

Among the most interesting features of this area are the gaps by which the Weald rivers escape northwards through the Greensand and Chalk ridges into the London basin or the Thames estuary, and southwards through the Chalk into the sea. At each of these gaps, with the exception of that of the Cuckmere, is a town, the sites being somewhat similar to that of Lincoln, marking the places where the east–west routes were compelled to descend to cross the rivers, and where modern routes following the valleys now cross the ridges. Guildford on the Wey, Leatherhead and Dorking on the Mole, Maidstone and Rochester on the Medway, Lewes on the Ouse, and Arundel on the Arun all owe their sites to similar conditions: many of them have old castles guarding the ancient routes. Lewes, which Defoe described as 'in the most romantic situation I ever saw', is at a confluence where a minor fold more or less detaches the upland of harder Middle and Upper Chalk on which the town stands. Arundel is also striking as a meeting place of routes; here the Chalk gives rise to steep river cliffs. Guildford is situated in the Hog's Back area where the chalk outcrop is very narrow, and in consequence it stretches through the Wey Gap, really uniting two earlier settlements, one to the north and one to the south of the Chalk.

The origin of these gaps probably calls for little comment, since they are in many respects similar to those at Lincoln (p. 40) and at Goring (p. 60). The rivers flowing northwards from the High Weald rise on the sandstone hills and cross a clay belt before cutting their gaps in the Greensand ridge, after which they cross another clay vale before cutting gaps in the Chalk. Throughout the greater part of their courses they are flowing with the dip, and though they are joined by strike streams flowing along the clay belts of the Vales of Kent or Holmesdale, where some themselves become meandering strike streams, their essential character is clearly related to the dip of the beds, and they may be called consequent streams. The cutting of the gaps was obviously begun before the clay belts had been reduced below the level of the uplands, when the rivers flowed along a dipping surface inclined to the north or south. The present relief has thus been produced while the gaps have been steadily lowered by the more rapid removal of the soft beds along the length of their outcrop, assisted by the growth of subsequent streams.

As these changes have taken place there have been many minor readjustments of drainage owing to some rivers having been able, by virtue of more advantageous positions or of greater cutting power, to erode more quickly than their neighbours; so their tributaries have cut back in the clay belts in a way that enabled them to capture the headwaters of their neighbours. Thus the Medway has been able to behead the Stour and Darent, the water from the High Weald formerly flowing into these rivers being now transferred to the Medway by east and west tributaries flowing in the Vale of Kent. Similarly the Wey has grown at the expense of the Blackwater, which formerly rose to the south of the Chalk; the headwaters of the present Wey belonged to the Blackwater but are now carried along the Gault Clay outcrop eastwards from Alton and Farnham to Guildford. Here in fact the Wey is draining some of the chalk plateau of Hampshire beyond the limit of the Weald, and it is also worthy of note that the water draining northwards from the Weald finds its way into the London basin and the Thames or the Thames estuary. The strengthened Wey has thus cut for itself a deep gorge-like valley near Godalming but the reduced Blackwater now has a valley too large for its present stream, which may be called a 'misfit'. All these changes, however, have tended to make the clay vales lower and wider, and to bring the harder out-crops into stronger relief, while making the more extensive clay tracts increasingly monotonous.

But if the rivers of the northern part of this area originated on an inclined surface sloping to the north, it is equally true that the rivers of the southern part originated on a similarly inclined surface sloping to the south. The Arun, the Adur, the Ouse and Cuckmere are also mainly dip streams, with subsequent streams along the wide Vale of Sussex (notably the West Rother). It follows then that the rivers of the Wealden area originated on a surface which was folded to north and south, more or less as the rocks are folded at present. The relationship of the rivers may more easily be under-stood if they are supposed to have begun when the Chalk extended right over the area, though as a matter of fact some of the upper beds of the anticline had already been removed before the rivers began to erode the surface.

The rivers of the Weald may thus be quoted as an example of close adjustment of drainage to structure. Most of them are subse-quent or strike streams flowing along the transverse clay vales for the greater part of their length. This is not surprising since most of the Weald and parts also of the Downs have not been submerged since Eocene times.

The Cliffs and Marshes of the Coast

The coastal scenery of this area, like the surface relief, is intimately related to the geological structure. The chalk cliffs of Dover have already been referred to; these continue northwards to Deal where the cliffs fall away as the Tertiary rocks are reached. Sands and clays of this series form the wide bay at Sandwich where the Stour seeks so long for an exit to the sea, while to the north the Chalk reappears in the cliffs of the Isle of Thanet at Ramsgate and Mar-gate, brought up as a result of a small fold and separated from the Downs by the syncline in which Canterbury lies.

The Isle of Thanet was formerly separated from the mainland by the open Wantsum channel, now entirely silted up. In Roman times this channel was guarded by the forts at Richborough Castle and Reculver. A shingle spit grew from the north and deflected the outlet of the river Stour. The port of Stonar, prosperous in medieval times, was built on the end of this spit. Another spit grew north and deflected the Stour north again. Sandwich was built on the opposite bank of the river to Stonar. To the west of Reculver are the low London Clay cliffs of the north Kent coast, at Herne Bay, Whitstable, Sheppey and the Isle of Grain, all subject to erosion where unprotected.

Returning to Dover, and following the coast around to the south, at Folkestone the Gault Clay forms the foreshore to the east of the town, and the cliffs of dull white Chalk rest on this clay foundation. As a result, water passing down through the Chalk (and the Upper Greensand just beneath it) is held up at the top of the blue clays, and along the bottom of the cliff gives rise to a line of springs, the flow of water keeping the clay always wet and slippery. These form ideal conditions for the development of landslips in the jumbled and untidy face of the cliffs (Plate 10). At Dover, on the other hand, the blue clay is far below sea-level and the cliffs remain strong and present clean, regular surfaces to the sea.

The west cliffs of Folkestone differ strikingly from those of the east, for there is no Chalk, the beds below the Chalk being here exposed. The Lower Greensand gives rise to a series of low and variable cliffs, sometimes steep where hard sandstones break along vertical joints, through Folkestone to Hythe and Sandgate. The cliffs gently sink to the level of Romney Marsh, where the Vale of Kent meets the coast. This flat tract, now largely drained and used for sheep farming, is traversed by weedy streams which meander through the low ground as they seek a path to the sea, lacking energy to carry away the material they bring down, and ready to deposit it and to extend the flats. For the marsh is made up of alluvial material, fronted along the coast by diverging ridges of shingle, culminating in Dungeness, the largest accumulation of shingle in Britain. Behind the marsh the land rises sharply in what may well be an old line of cliffs, which is very noticeable continuing inland westwards from Hythe to near Appledore.

Probably at an early stage in its development the area of the

Plate 10. *Folkestone Warren, Kent. In the background are cliffs of Middle and Lower Chalk, which overlook slipped and fallen masses of Chalk and Gault. The slipped material overlies the Gault outcrop. Note the sea defences, and the railway which runs through the landslip area.*

present marsh was occupied by a wide bay, now represented by low ground from Winchelsea to Hythe; the sea at that time attacked the coast and cut the low cliffs which form the northern boundary. This bay is believed to have come into existence by the drowning of the lower part of the Rother valley, already eroded deeply along the soft belt of the Weald clays. The drift of shingle along the coast from the south-west, and probably small changes in sea-level, may have led to the formation of a spit partly or wholly across the bay, and so to the impeding of the rivers and the silting up of the bay by the deposition of river-borne material. There have been many changes in the shingle beaches, and the history of Dungeness is very complicated, a large series of curved banks between Lydd and the shore marking former positions of the coastline. At one time the Rother reached the sea at New Romney, on the north of Dungeness, but it now cuts through the shingle some eight miles farther west. More recent changes in this piece of coast have left their mark on its history: the destruction of the old town of Winchelsea by the advance of the waves led to the designing of New Winchelsea as a channel port by Edward I, but the sea treated this in a different way, leaving it some way inland. Henry VIII, who sought to defend the estuary by a castle, was no more fortunate, for in a short time his Camber Castle was also left some way inland. The deep water off the point of Dungeness has led to the building of lighthouses, coastguard stations, and two nuclear power stations, with their huge demands for cooling water. Something is thus known of the development of Dungeness and the marshes it encloses, but there is no real answer to the problem of why shingle should accumulate in this particular area. This awaits further research.

Between Winchelsea and another area of marsh at Pevensey Levels is a length of cliffs in the Hastings Beds of the High Weald, at Fairlight and Hastings. Through Bexhill and Cooden these cliffs become lower and give way to the marsh coast of Pevensey Levels, which bears some resemblance to Romney Marsh, but the shingle formations are almost insignificant compared to Dungeness, Langney Point being the shingle foreland here. Changes are shown by the Roman and Norman Pevensey Castle now standing inland.

Beyond the Pevensey Levels Eastbourne shelters under the chalk headland of Beachy Head, whence high cliffs extend through Brighton and with slight interruption almost to Bognor. At many places the cliffs are vertical, and are deeply cut along joint planes to form caves and projecting stacks of peculiar forms. Beachy Head, with its cliffs over 500 feet high, represents the eastward termination of the South Downs, which are there cut across at right angles; thence to near Worthing the coast follows the trend or strike of the beds and the cliffs retain a general uniformity of type for many miles. Here the cliffs truncate the dip valleys which formerly extended farther to the south but which have been shortened by the attack of the waves on the cliffs, the coastline moving steadily inland (Fig. 12). The cliffs thus vary in height according to the relief of the land they limit, as in the Seven Sisters, near Eastbourne.

The coast of south-eastern England thus presents two different aspects: from Margate to Beachy Head, alternating ridges and valleys made by hard and soft rocks are cut nearly at right angles by the coastline, and there is a corresponding alternation of cliff and wide beach or marsh. South-west of this the coast more uniformly follows the trend of a single rock group, the Chalk, and its variety there depends on the relief and details of structure.

7. The South Coast

The region dealt with in this chapter is extensive, stretching along the south coast from near Worthing westwards to the borders of Devon; thus the region abuts on to the areas described in the chapters on the Weald, the chalklands and south Devon. Much of it constitutes the Hampshire basin, but a large part of it, the Isle of Wight, is detached, and another important part, the 'Isle' of Purbeck, is nearly detached, from the mainland. The Chalk forms the coast along part of the area, both to east and west, while in the Isle of Wight and to the west rocks older than the Chalk are also found. It is thus an area of some geological complexity, although in many respects it resembles a partly flooded London basin. Much of the area of south Hampshire and east Dorset is a syncline holding Tertiary rocks. Chalk surrounds these Tertiary rocks, although to the south it is breached between Swanage and the Needles and again between the eastern end of the Isle of Wight and Bognor Regis, allowing the sea to fill Bournemouth Bay, the Solent, Spithead and Southampton Water. The southern border of this basin has more variety of structure than in the London basin and its coastal scenery is of extraordinary interest.

Hampshire and East Dorset – the Tertiary Beds

A convenient starting point for the description of this region is the central area covered by the youngest rocks found here – the Tertiary beds – which overlie the Chalk. The Tertiary beds cover roughly an area south of a line joining Worthing, Chichester, Romsey, Fordingbridge, Wimborne Minster, Dorchester and Swanage, as well as the northern part of the Isle of Wight. which will be dealt with separately, thus including parts of Sussex, south Hampshire and east Dorset. All the towns named except the first and last, which are primarily seaside towns, are examples of the more important towns frequently found along a geological boundary.

The Tertiary beds cover a great area of rather low relief. As in the London basin these rocks consist of sands and clays, but in this area there is a greater thickness and a more varied succession, while on the whole there is a bigger proportion of sand. So that whereas in the London basin the clays (especially the London Clay) make up vast stretches of heavy country, in this area there are greater expanses of drier sandy country comparable, for example, with that north of Aldershot, and illustrated in Plate 7. Over these sandy areas extend great commons and woodlands, of which the New Forest is a notable example.

It is probable that trees disappeared from Britain entirely during the Ice Age, for although the ice did not come farther south than the Thames basin, the country which was not ice-covered was so cold that it could maintain only a tundra vegetation of hardy arctic and sub-arctic plants, similar to those which can still be found in Britain today on Cairngorm summits in Scotland. As the ice receded and the climate improved, forests spread northwards, the birch and pine (which are still the dominant trees in Europe to the south of the tundra belt) being among the first to enter England. These were followed in time by the oak, ash, elm and hazel. In the face of this competition the pine shortly afterwards became extinct, and was not again found in England until its introduction by man in the last three centuries, since which time it has spread rapidly, often at the expense of its competitors. Pines are now spreading over many areas where deciduous woodlands formerly flourished. The pine-woods of the Bournemouth area exemplify the success with which this tree has colonized these sandy areas, while oak, ash and hazel remain on the damper clay soils. In the pine-woods there is little opportunity for undergrowth, for the thick carpet of pine needles prevents the growth of many plants. But in the more open parts of these forests bracken is abundant.

The forest-lands give to this area its most distinctive character, for in this part of England and in the western region, woodlands cover a bigger proportion of the land than elsewhere, in spite of the great quantity of oak cut for shipbuilding, which continued in the estuaries of the Hampshire coast until about a century ago. In many parts of the country farms still appear as isolated clearings in the forest, and the villages scattered over the area are mostly small, at any rate those away from the coast, and are connected by roads which seem to straggle aimlessly across the countryside; probably many of them representing forest tracks.

The present distinctive appearance of the New Forest owes much to its peculiar legal status. Although not continuous woodland at the time, it was brought under Forest Law by William I for his pleasures of hunting and it has remained Crown land ever since. Thus, although subject to encroachment from time to time, it has remained an area of open heath and woodland, available for recreational use in modern times, together with forestry enclosures. Minor east–west folds intrude small areas of Chalk into this Tertiary tract. One such area is Dean Hill, affording fine views over the valley to the north, floored with Tertiary rocks, and containing the villages of East and West Dean (Fig. 8).

East of the New Forest, the Tertiary area is narrower, and near Fareham and Havant the structure is complicated by other smaller folds which result in an area of Chalk being brought up into the Tertiary lowlands. This isolated tract of downland, known as Portsdown, is some ten miles from west to east and rarely a mile across. Between this and the wider chalklands to the north lies a low belt of Tertiary beds, about three miles wide, occupied by rich woodland, including the Forest of Bere. London Clay forms the greater part of this belt, but small patches of Bagshot and other sands in the deeper parts of the fold give rise to the irregular heaths of Purbrook and Walton. Chalk also outcrops near the coast between Bognor Regis and Worthing, but without surface expression. This lowland and that around Selsey Bill were covered by a shallow sea late in the Ice Age; old cliffs can in places be seen at the back of the plain. It is marginally higher than the area which at present contains Chichester, Langstone and Portsmouth harbours.

The deeper channels in these harbours, as well as, far more obviously, Southampton Water, represent former river valleys cut at other episodes of the Ice Age when sea-level was, relative to the land, appreciably lower than at present, and the rivers flowed away to Spithead, which was itself a former river valley. The former drainage of the Hampshire basin as a whole is dealt with below (Fig. 13).

At the western end of the Hampshire basin, west of the New Forest, there is again a bigger proportion of sandy beds, and many heaths and some woodlands remain, unused for modern agriculture. Some of these heaths in Dorset, as well as some beautiful country in the Isle of Purbeck, are used as army training areas.

Here the Hampshire basin narrows considerably and its direction is marked by the eastward-flowing river Frome, on which Dorchester is situated at the Roman crossing. Here tributary valleys, coinciding with the dip of the Chalk, trench deeply into the upland and cut it into narrow north–south ridges along which ancient ridgeways have been converted into modern roads. In this region the Frome is running along the axis of the synclinal fold, the relationship of this river and its tributaries to the basin being therefore

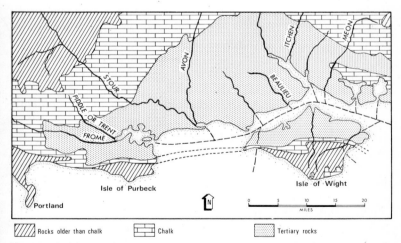

Fig. 13. *The Hampshire basin, Isle of Wight and Isle of Purbeck, showing the geology and rivers, particularly the former continuations of the Chalk and the Frome–Solent river.*

very closely comparable with those noted in the London basin. It appears that the Frome was developed along the bottom of a fold which was tilted to the east, and that into it dip streams flowed from both north and south. Its course to Wareham and into the sea below Poole accords with this interpretation.

Still farther east the Solent and Spithead continue along what is practically the same line, and it is considered that formerly the Frome extended eastwards along this line, receiving its tributaries the Stour, Avon and Test on its left bank and the Isle of Wight rivers on its right bank (Fig. 13). Of course at this time there was much land to the south of the present shores, and Bournemouth Bay and the area to the south of it must have been land; possibly other tributaries from the south joined the Frome in this region and began the cutting of the wider breach in the Chalk there. The similarity of this extended Frome, flowing along the bottom of a shallow trough, to the Kennet and Thames in the London syncline needs no emphasis, the tributaries in both cases following the dip of the beds. But in the case of the Hampshire basin the whole river system has been drowned under the sea, as a result of the lowering of the level of the land or rise of sea-level in post-glacial times referred to already. In this way Southampton Water was formed by the flooding of the lower valley of the Test, and a deep-water approach is possible to the port of Southampton. This port enjoys a further advantage over London and Liverpool, in that the area of its approach channels has not been glaciated, and does not require continuous and extensive dredging to remove the debris of glaciation.

Wave action has also greatly modified the remnants of this extensive river system, for besides cutting into the south coast of the Isle of Wight and beheading the rivers there, the waves have cut deeply into the mainland on the northern side of the original valley, on the coast of Bournemouth Bay; near Bournemouth they have cut low cliffs in the sands and clays. Along this coast some of the streams in the narrow steep-sided chines may constitute the last remnants of rivers formerly tributary to the Frome, whose lower courses have been removed by the advancing sea. In Branksome Chine a narrow gorge has been cut in the floor of a much wider valley, and it may be suggested that this latter represents the old tributary valley, the cutting of the gorge within it having been brought about by the steepened gradient given to the stream by the advance of the sea.

It may be worth pointing out the contrast between the effects of marine erosion on the rivers of the Isle of Wight and those of the Bournemouth area: in the former region the sea has cut away the narrow steep-sided valleys characteristic of the upper part of the river courses, leaving the stream with a very gentle gradient, while in the latter the sea, advancing towards the source, has left only the upper parts of the valleys, but with an increased gradient.

Bournemouth Bay exhibits several other interesting features. It is formed between two headlands of more resistant rocks: the Chalk of the Foreland, ending in Old Harry Rocks, and Hengistbury Head, formed of ironstones in the Bracklesham beds. Between these two headlands lie the low sandy cliffs of Bournemouth, and the drowned lower reaches of the river Frome in Poole Harbour, protected from the open sea by twin spits pointing in opposite directions. Of these two spits, the South Haven peninsula to the south near Studland appears to be growing by the addition of sand, while the Sandbanks peninsula to the north appears to be temporarily at least fixed, even if walls and groynes are required to keep it so. Such double spit features pose many interesting problems to students of coastal processes, and the pattern of Bournemouth Bay is repeated to some extent eastwards along the coast. Christchurch Bay has its harbour, protected by a double spit, and its eroding cliffs, while at its eastern end is the fine shingle spit of Hurst Castle, extending out into the Solent. Pagham Harbour beyond Selsey Bill is also guarded by a double spit.

The Isle of Wight

The Isle of Wight is the next convenient division of this south coast region for description. Cut off from Hampshire by the narrow waters of the Solent and Spithead, it has escaped some of the development which has affected many parts of the south-east coast, while its isolation has given greater attraction to its quiet beauty.

The most important geological feature in the island is that ridge of
Chalk which runs in a slight dog's-leg from the Needles in the west
to dazzling white Culver Cliff on the east, a narrow low ridge of
downland which divides the island into two areas of entirely different character, and which constitutes the long diagonal of its
lozenge shape (Figs. 8, 13 and 14).

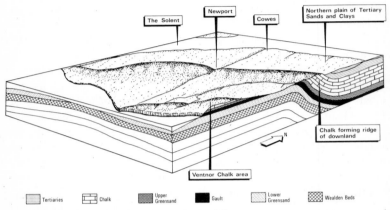

Fig. 14. *Block diagram of part of the Isle of Wight.*

This Chalk outcrop is for the most part narrow, for the beds here
are dipping steeply and are often nearly vertical. The ridge is thus
closely comparable with the Hog's Back which in the Guildford
area forms the boundary to the London basin: the hog's-back
pattern is more persistent in the south of the Hampshire basin.
The chalkland across the middle of the island exceeds 500 feet in
height in only three areas, but it stands out prominently above the
clays and sands which flank it both to the north and south. Near
the middle of the island, south and west of Carisbrooke, the dip of
the Chalk is lower and it spreads over a much wider area, about
three miles across: here the downs are higher, Brightstone Downs
reaching 700 feet, and are deeply dissected by narrow dry valleys.
To the north of the chalk ridge the island is made up of Tertiary
strata which compare with those across the straits on the mainland,

while to the south clays and sands like those of the Weald form much of the island.

The northern part of the island, from the chalk ridge to the Solent and Spithead, is for the most part low-lying, and largely made up of clays, as opposed to the largely sandy nature of the Tertiary areas on the mainland. The soils are heavy and there are many copses and woods, of which Parkhurst Forest is the most extensive. Along the Spithead coast, these soft rocks generally give rise to a low, shelving shore-line, the narrow sea looking like a great river in a wide, shallow valley, and when low cliffs are present they are often overgrown. But along the Solent the coast is sometimes bolder, especially where thin limestones occur among the clays. At the eastern extremity of the island, the Foreland is made up by the Bembridge limestone, a harder bed occurring among the usually soft Tertiary strata. Where so few stones can be used for building, it is not surprising that this limestone has been used extensively, for instance in Yarmouth Castle, and it has also been exported into the Hampshire region. Elsewhere in this northern area bricks have been made from local clays, and they have been used in the majority of smaller buildings.

The wide spread of the Tertiary rocks as compared with the narrow strip of chalk outcrop is noteworthy: the Chalk is more than 1,500 feet in thickness and forms a belt only half a mile wide in places, whereas the Tertiary beds are about 2,000 feet thick and occupy the whole of the northern tract, in the centre five miles wide. These differences in distribution are due to the structure; while the Chalk is for the most part nearly vertical along this belt, so that the width of outcrop is not much greater than the total thickness, the Tertiary beds, except just along the edge of the Chalk, dip very gently, and in many parts are nearly horizontal (Fig. 14). The Chalk and the rocks near it plunge steeply down in the middle of the island but almost at once curve away gently to rise to the north. the Chalk reaching the surface again near Winchester and on Salisbury Plain. The basin thus produced is markedly asymmetrical. and its deepest part lies much nearer the south than the north. It may be noticed that the London basin shows a similar asymmetry. its steepest dips (as in the Hog's Back) being likewise in the south.

The more interesting part of the island, however, lies to the south of the chalk ridge, where rocks belonging to the same series as those in the Weald form a broad belt: the Greensands, the Gault Clay, the Weald Clay and Sands are all represented and are seen in the low, yellow-green cliffs between Sandown and Shanklin. Accordingly this region shows greater variety of scenery than that of the north. In the main these more readily denuded beds form a wide vale, extending from Brightstone in the south-west to Sandown in the east, with fertile, red-brown soils and rich green fields, a region of gentle relief bounded by wooded ridges made by the sandy beds, above which rise both to north and south the downlands made by the Chalk.

For this southern vale is essentially an anticline, more or less comparable with the Weald, but with generally steeper dips on the north and gentler on the south. Thus in the most southerly corner an outlier of Chalk introduces a delightful tract of downland, rising in St Boniface Down to nearly 800 feet and forming the highest land in the isle; this wide outcrop of Chalk, with its greater elevation and the steep scarp of its northern face, contrasts very sharply with the character of the central ridge. These southern downs are deeply indented by small valleys on the northern scarp, but near Ventnor on the south coast the porous Chalk and Upper Greensand overlie the 'blue slipper' formed by the wet Gault Clay. Here there is a steep, well wooded slope much disturbed by landslips, with the conditions described at Folkestone (Plate 10) repeated on a grand scale. As the whole of the rocks have a gentle, southerly dip, there is a strong tendency to slip seawards, and damage to the coast road beneath the cliff has been a frequent occurrence.

Although the country between the southern downs and the central ridge is generally low and has been spoken of as a vale, it is not related to the main drainage of the island, which presents many features of extraordinary interest. All the more important rivers — the river Medina and the Western and Eastern Yar — rise in the southern area, not far from the coast, and flow into the Solent and Spithead after cutting gaps through the chalk ridge, while further 'wind gaps' are found marking former stream courses.

The courses of these streams are related to the northerly dip of the rocks over most of the island, and were determined when the area emerged from an early Pleistocene sea, long before the central vale was eroded to its present form, the rivers having maintained their ancient lines. The situation is more or less similar to that described in the Weald and Wessex. The situations of Newport and Brading in these gaps need no comment. But the positions of the sources of the rivers are very remarkable. The Western Yar rises very near the coast about three miles from the Needles, and the upper part of its valley cuts right into the cliffs at Freshwater Bay, so that a very little depression of the land would make this western area into a separate island. Moreover, the gravels laid down in this valley are found right on the south coast and extend to the Solent: obviously the river which cut the valley and deposited the gravels must at one time have been longer, and its headwaters must have risen some distance to the south of the present shore-line. From this vanished land it must have obtained the material which was deposited in the tract which then formed the lower part of its course. It is certain that wave action has led to the destruction of quite considerable areas of land which lay to the south of the present coast, and that this destruction has occurred since the rivers had carved their valleys to approximately the present forms. The river courses must further have extended from their present mouths to join in former times the Frome–Solent–Spithead river as south-bank tributaries (Fig. 13).

The Eastern Yar has similarly lost much of its drainage area by the encroachment of the sea, the valley of a tributary near Sandown showing evidence of recent truncation. On the other hand the Medina, rising in the southern chalk hills, has not so far suffered by this marine advance.

Apart from this essentially northerly drainage, undoubtedly the original drainage of the area, there is little tendency to develop new streams except as tributaries subsequent to the main dip streams. But here and there small streams descend rapidly from the higher ground near the south coast into the sea. Such streams have a steeper gradient than any of the main rivers and they are able to cut small, deep ravines such as that of Shanklin Chine, carved in the Greensand beds. or Blackgang Chine on the south-west coast.

Apart from this southern portion, the most interesting points of the coast are at the two extremities of the island, where owing to the locally steepened dip many different beds are brought into the coastline in a short distance, and give rise to coastal scenery of rare variety of colour and form. We have already noticed that here as elsewhere the Chalk tends to form headlands. The Needles at the western end are familiar; here wave action has left small pointed islets of Chalk by cutting through the ridge. Standing on the chalk cliff above the Needles, they are seen to fall into line with the main ridge, and on a clear day, looking along this line, the cliffs of Chalk under Ballard Down in the Isle of Purbeck, across Bournemouth Bay, may be seen continuing the ridge. The former connection of these areas is then not so difficult to appreciate, and just as the sea has detached the Needles from the island, so the wider breach may have been created, though over a longer period of time (Fig. 13).

South and east of the Needles the waves have worked away the rocks which underlie the Chalk, and for some miles the cliffs expose the southern face of the ridge, forming steep, vertical faces of as much as 400 feet. But to the north of the Needles the Chalk is overlain by the sands and clays of the lower beds of the Tertiary, whose steeply dipping beds prove more easily eroded than the Chalk, and have been excavated to form Alum Bay, where the variegated sands and clays make a pleasant contrast to the white cliffs of the Chalk. Whitecliff Bay at the eastern end of the island occupies a precisely similar position.

The Isle of Purbeck

The Isle of Purbeck contains, in its chalk 'spine', a continuation of the central chalk ridge of the Isle of Wight just described. This is a hog's back ridge of nearly vertically dipping Chalk, comprising part of the southern chalk rim of the Hampshire basin. From the top of this chalk ridge, the sandy heaths of the Tertiary deposits can be seen stretching away to the north, while to the south lies a near repetition of the scenery of the southern half of the Isle of Wight, and the features of the Isle afford some support to the views regarding the origin of the scenic features already discussed.

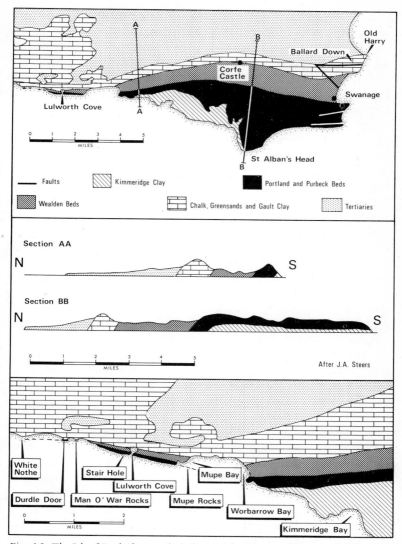

Fig. 15. *The Isle of Purbeck. A geological map* (above), *two sections* (middle), *and a more detailed map of the Lulworth Cove area* (below).

Key (map above):

- Faults
- Wealden Beds
- Kimmeridge Clay
- Chalk, Greensands and Gault Clay
- Portland and Purbeck Beds
- Tertiaries

Labels (map above): Old Harry, Ballard Down, Corfe Castle, Swanage, Lulworth Cove, St Alban's Head, A—A, B—B

Section AA — N ... S

Section BB — N ... S

After J.A. Steers

Labels (map below): White Nothe, Stair Hole, Lulworth Cove, Mupe Bay, Durdle Door, Man O' War Rocks, Mupe Rocks, Worbarrow Bay, Kimmeridge Bay

MILES 0 1 2 3 4 5

Of course the Isle of Purbeck is no island, but it used to be more of an island than it is now, for Poole Harbour was formerly flanked by marshes which extended far along the Frome valley. This can well be visualized at Wareham, where from the north the road climbs a rise into the Saxon town, and to the south a former causeway connects the town with Stoborough. This tract of country, some fifteen miles long and seven or eight miles wide at its maximum, is bounded by cliffs to east and south, but it tapers westwards, and in that direction it joins so unobtrusively to the 'mainland' that its boundary is not well defined. The Isle is also an area of attractive coast scenery which merits very special consideration (Fig. 15).

We have already pointed out, from a viewpoint above the Needles, the continuation of the chalk ridge westwards into Purbeck. On the east of Purbeck the typical chalk uplands of Ballard Down end in vertical white cliffs, the lower part deeply fretted by wave action, which has detached numerous stacks, including the famous Old Harry Rocks. These remains of the old cliff face show that here, as in the Isle of Wight, the coast is receding slowly under the attack of the sea. The chalk uplands form a very narrow tract, often not more than a quarter of a mile wide, once more the result of steeply dipping beds which in many places are quite vertical. As in the Isle of Wight the drainage flows here from south to north, to join the Frome, and the downland is cut by a narrow gap at Corfe where the grey stone village is overtopped by the large castle which guarded the converging routes.

North of the downs occur rocks like those of the Hampshire basin, sands and clays which have been worn back by the sea more rapidly than the Chalk of the headland. Studland Bay is thus closely comparable in structure and scenery with Alum Bay and Whitecliff Bay; here instead of chalk cliffs dropping sheer into deep water we have a gently sloping sandy beach backed by attractive cliffs on which vegetation soon becomes established. To the south of the chalk ridge the rocks are at first similar to those of the Isle of Wight, the most important group, the Wealden beds, being mainly clays which have been worn back to form the beautiful Swanage Bay (Fig. 15). Nevertheless the country formed by the

Weald clays is mostly lower than that to the north and south, and it makes up a belt at first over a mile wide, but tapering gradually westwards to Worbarrow Bay owing to the diminishing thickness of the clays. The thinning of these beds, while the Chalk continues unchanged, will be understood when their origin in a lake of limited extent is borne in mind (p. 94); the shores of the lake were situated not far to the west of the Isle of Purbeck.

South of this clay belt the land surface is once more high, and where it meets the coast it forms the headlands of Peveril Point and Durlston Head. The east coast of Purbeck thus presents an alternation of projecting headlands backed by high ground and shallow bays fronting tracts of low ground. The bays are found along beds which are easily eroded, whether by the sea, or by rivers and the atmospheric agencies of frost and rain; the headlands consist of harder and more resistant strata. These conditions of alternating bays and projections are often found along coasts where the sea is succeeding in its advance, or where a ridged land surface has been partly submerged.

The south-eastern part of Purbeck is made up chiefly of limestones, which are more gently inclined than the rocks of the middle of the 'island': in many places along the coast they dip gently seawards, and the general structure closely parallels that of the south of the Isle of Wight. But as the fold rises westwards older beds are exposed along the coast, and the rocks seen on the south of Purbeck are not seen at the surface anywhere to the east. The limestones are of various kinds: immediately under the Wealden beds are limestones of freshwater origin, rich in the shells of freshwater snails, and they were formed in a forerunner of the Wealden lake. These beds are known as the Purbeck Limestones: they are tough and resistant, and have been widely used for building, some being specially important because they will take a good polish, for which they are commonly known as marbles. They have been very extensively employed in the interior decoration of churches throughout southern Britain.

Among these limestones are softer groups of clays which give rise to the shallow Durlston Bay, but, generally speaking, these beds and the Portland stone which underlies them create an impressive line of cliffs which reaches west to St Albans Head, and terminates a plateau averaging about 400 feet above the sea. The limestones being nearly horizontal, the joints which cut the beds at right angles are almost vertical, and these have largely determined the form of the cliffs, which are steep and incised deeply along the weaker beds.

Scattered on this limestone plateau are the many old quarries where the stone has been dug, and many small hamlets of grey stone. But the area is lightly peopled, for the soil is not very productive and there is little connection between the villages and the sea, for fishing has not prospered along this coast as it has farther west. Thus few houses are situated near the shores, and from many of the villages placed above this rocky coast there is no good road to the sea. Beyond St Albans Head, however, the cliffs gradually fall away as the Kimmeridge Clay rises from beneath the sea-level. At first this clay outcrop forms the base of the cliffs, and here, along the west of St Albans Head, landslips have given rise to an irregular coastline, the thick limestones having slipped over the impermeable clays owing to the gentle, seaward dip. Still farther west the clays form the coast in Kimmeridge Bay.

Lulworth and the Coast to the West

It is, however, along the coast some five miles west of this that the coast scenery is most fascinating, in the stretch of shore-line which includes Worbarrow Bay, Lulworth Cove and Man o' War Cove (Fig. 15). Throughout most of this tract the coastline follows the strike or trend of the beds, and shows a variety of coastal form resulting from the alternating harder and softer rocks, in contrast to the condition in the eastern end of Purbeck, where the cliffs cut right across the various beds. Here in the west the position of the coast is determined mainly by the harder limestones of the Portland beds which form an almost unbroken line of rugged cliffs. These beds are steeply inclined to the north, and the underside of the rocks forms in many places an overhanging cliff, the softer Kimmeridge Clay having been stripped away for most of the distance. Along this part of the coast there is no tendency to landslips, for the

Plate 11. *Stair Hole, Dorset, with Lulworth Cove in the background. The acute folding and steep dip of the Portland and Purbeck beds can be seen to the right, both in Stair Hole and on the far side of Lulworth Cove. The Wealden beds form the lower ground to the left of Stair Hole and beyond Lulworth Cove. The Chalk rises steeply inland of Lulworth Cove.*

beds are dipping steeply inland, and so there is thus little tendency for them to slip seawards.

Along almost the whole of this stretch of coast, therefore, the waves are hammering at the steep bedding plane of limestone, and in a few places they have succeeded in cutting a way through the limestone group, which forms only a narrow belt, and on reaching the softer group beyond have commenced the excavation of a larger bay. This process has left examples surviving in different stages at different places along this coast, at the same time, and thus it can be beautifully illustrated.

The best starting point for this illustration is Stair Hole, just to the west of Lulworth (Plate 11), where two small holes cut near the base of the limestone barrier have allowed the sea to scoop out deep holes in the softer beds behind. The next stage is Lulworth Cove itself, with its nearly circular plan and narrow entrance, offering one of the few safe anchorages along this rocky coast. Next consider Worbarrow Bay and Mupe Bay, probably two Lulworth-type coves which are wider open and have joined to form one wide bay, with Mupe Rocks at the western end, and a line of rock stacks continuing the line of the limestone barrier. This shows how the barrier is only gradually being broken down, the waves meanwhile advancing more rapidly to destroy the softer rocks behind. The next stage is seen in the Man o' War Cove area, where only fragments of the barrier survive as at Durdle Door with its arch, Man o' War Rocks, the Cow and Calf and others. Finally, to the west the barrier has disappeared and straight chalk cliffs are seen near Bat's Head. The situation is not quite as simple as these stages suggest, with complications involving the changing dip of the rocks, and possible aid to cove formation by streams working in the soft Weald Clay group, but it illustrates graphically the possibility of various stages existing simultaneously, in what might have been thought of as a cyclic evolution. This area also illustrates on a small scale the breach made in resistant rocks between the Isles of Purbeck and Wight.

West of the chalk cliffs of White Nothe, the coastal rocks are again older than the Chalk. The Purbeck-type structure is repeated with variation in the Weymouth area. An anticline, complicated by

Plate 12. *Chesil Beach, Dorset, looking north-west from near Weymouth. The curve of the shingle bank is clearly shown, separating the sea from the Fleet, the 'lagoon' behind the bank. The rocks inland of the Fleet are a variety of Jurassic beds, and the curve of Lyme Bay is seen in the background.*

faulting, to the south of the Chalk around Dorchester, brings rocks as old as the oolites and the Oxford Clay to the surface, and Weymouth Bay is cut mainly in Oxford Clay. The southern limb of this anticline dips gently, with Portland and Purbeck beds at the surface. These beds with their gentle dip are responsible for the distant views of the Isle of Portland as a surface tilted gently southwards.

The Isle of Portland has been so scarred by quarrying and in the main is so treeless and drab, that except as a viewpoint it has little attraction. The view from the heights of Portland, however, is magnificent, of the pebble banks which link the isle to the mainland. A small bank carrying the road links Portland to Weymouth, while a much larger bank, the famous Chesil Beach, fronts the coast to the west for some sixteen miles to West Bay (Plate 12). It is not directly connected to the mainland for the first eight miles: along this stretch it encloses the tidal water of the Fleet, which is connected with the sea through Small Mouth, a breach in the small Portland–Weymouth bank. Chesil Beach is made up largely of rounded pebbles of flint (derived primarily from Chalk but probably reworked in Eocene and later times) and also of quartzite (derived from the New Red Sandstone pebble beds still further to the west) and also some local limestone pebbles at the Portland end. The pebbles are graded from about pea size at West Bay to fist size and larger at Portland. The accumulation of such vast quantities of material far from a direct source, as well as the extraordinary grading, has provoked much controversy and speculation as to the origin and mode of formation of the beach. Present thinking favours the accumulation of the material as a barrier beach during the post-glacial rise in sea-level, but many problems are not yet satisfactorily explained. The material is held up by the cliffs of Portland which, acting as a gigantic groyne, have checked further eastward movement. Hence the Portland–Weymouth pebble bank is smaller and composed largely of shingle derived from the local limestones.

If we go farther west we leave behind the rocks which give character to the Hampshire basin and to the coast of the middle part of southern England. But the cliff scenery is interesting right on into Devonshire. For many miles around Bridport, Charmouth and Lyme Regis the cliffs are cut in nearly horizontal rocks, limestones, and shales of blue and yellow, similar in many respects to those of the north Yorkshire coast and likewise forming steep cliffs, vertical where the limestones are strong and well jointed, sloping where there are thicker shales, with small reefs across the foreshore marking the position of the hard bands. Near Lyme Regis particularly the cliffs yield many fossils, the ammonites being particularly famous. Overlying these Lias rocks, Chalk and Greensand are found in the higher cliffs, occasionally slipped seawards, as in the famous Axmouth landslip, where in 1839 a great hollow 1,000 yards long was produced in one great slip, giving a tumbled mass of irregular mounds on the foreshore. Still farther west, near Seaton, the blue and yellow cliffs give place to cliffs of deep red, as the gentle easterly dip of the rocks brings up still older beds, for here the Lias is in turn underlain by New Red Sandstones, the colour of which gives a vivid richness to the cliff scenery at Sidmouth and Budleigh Salterton.

At Dawlish and Teignmouth these red rocks are often nearly horizontal. And when wave advance keeps the cliffs steep, the rocks are attacked along joint planes and give rise to the fantastic terra-cotta shapes which make the railway journey along this coast so attractive.

From here westwards the country changes its geological char-acter, from what may be broadly termed a continuation of the English scarplands, to the different geological province of Devon and Cornwall (the county boundary not following this geological boundary), which will be described in Chapter 21. From the Hampshire basin westwards, with the exception of the Purbeck and Weymouth anticlines, the coastal cliffs are cut in successively older rocks, which are of course also evident inland. The only major disturbance to this sequence is the presence, already noted, of Greensand and Chalk in the cliffs near Axmouth. These rocks lie nearly horizontally and unconformably on top of the older rocks, and inland the Greensand in particular has been planed off to give horizontal surfaces at about 800 feet north of Honiton, around Chard, and in the Haldon Hills south of Exeter (p. 69). In the Haldon Hills the Greensand is capped in its turn by gravels, mostly flint, derived from Chalk. Similar gravels are found around Hardy's Monument on Black Down in Dorset. It is thought that former, more extensive, spreads of these gravels may have been eroded away and eventually provided the present extensive flint accumula-tions such as Chesil Beach.

8. East Anglia

No part of England has greater individuality than East Anglia. Its towns and villages are distinctive, and there are no other extensive areas in Britain with such generally low relief. Yet East Anglia is mainly founded on the Chalk, which might be expected to give rise to such features as would link the area closely to the Chilterns which adjoin it on the south-west. East Anglia has, however, been covered by ice sheets during the Ice Age and over much of the area, therefore, the Chalk is buried beneath a cover of varying thickness consisting of glacial tills, sands and gravels; these superficial deposits to a great extent determine the nature of the soil, and their occurrence is responsible for many of the peculiarities of the area.

The most notable feature is the low relief, and in this it differs greatly from the other chalklands already noticed. For even along the western border of the chalk outcrop which extends through Suffolk and Norfolk, from near Newmarket to the coast at Hunstanton, its height exceeds 400 feet only at a few points in Suffolk, while in Norfolk it does not reach 300 feet; in Breckland it is below 200 feet, possibly because of the passage of ice sheets across it. In places this scarp rises sharply from the fenland on the west: from the scarp eastwards the chalk surface falls gently, and in many parts of the area Chalk is only seen in some of the deeper valleys, while at Yarmouth it is 500 feet below sea-level. It is therefore useful to regard the chalk surface of East Anglia as a gently inclined plane.

Part of this tilted surface, especially nearer the east coast, is covered by younger deposits of marine origin, known as 'Crag'. These represent the youngest marine rocks in England; in fact East Anglia may well be spoken of as one of the youngest parts of England, not long and not greatly raised from beneath the sea in which some of its deposits were formed. Its elevation has been reduced, in common with many parts of eastern and southern England, by still more recent subsidence, which has and is having important effects on the scenery. East Anglia has also been modified by glacial action.

In fact, notwithstanding the apparent simplicity of its solid rocks, it is a region of some complexity and includes areas of very different character. Its superficial rocks are of great importance and their interpretation has given rise to controversy; in this account it is not necessary to discuss these problems at any length, but reference is made to those conclusions which throw light on the origin of the scenic features.

The region has long been isolated from the rest of England. In early Neolithic times the dense forests of Essex on the south and the wide fenlands on the west almost completely cut it off from neighbouring populated areas; more recently the major routes from London to the north have been carried farther west in order to avoid the Fens and the Wash. True, the old land-route into the area, along the narrow strip of chalk downland from Newmarket towards Thetford, which carried the ancient track known as the Icknield Way, linked East Anglia with other populous chalklands to the south-west, but this connecting strip was so readily defended that the region long retained its independence and unity. At Domesday it was the most populous region of England, and its many villages and parish churches still testify to this early development.

Although this is an area of low relief, it is by no means a plain: parts of it are flat, especially the level alluvial stretches by the rivers and broads, but much of it is gently rolling country, often with rather surprisingly deep valleys. For the most part this undulating country is covered with arable fields, for two-thirds of the cultivated land in Norfolk is under the plough, and corn crops are very extensive owing to the dry climate and the light soils. Thus, over wide areas, pastoral lands are very restricted and the great stretches of cornfields give character to the scenery. This region has none of the dullness and lack of variety which sometimes characterize districts of low relief.

Although they are not very high, the chalk areas along the

western border of East Anglia are similar to the other chalklands of England. Around Newmarket there are rolling downs covered with short grass, with dry valleys, while in western Norfolk the Chalk runs to the coast at Hunstanton and gives rise to low cliffs. To the south of this, around Sandringham, a small area of Greensand, the rock next beneath the Chalk, is found, making a tract of pleasant heathland in which the Royal estate is situated. In some of these western chalk areas flint has been obtained from almost the beginning of human history, though the mining carried on so extensively at Grime's Graves, near Brandon, was chiefly done in the late Neolithic and Bronze Ages. In that area the chalk surface is marked by hundreds of saucer-shaped hollows, varying in diameter up to seventy feet, the sites of ancient mine shafts made in the working of particularly suitable bands of flint. On these hills to the northeast of Brandon it is possible to descend one of these old mines, and at Brandon itself to see the surviving flint-knappers at work. The use of flint, so long characteristic of East Anglia, has continued with its employment as a building material, and nowhere else in Britain has it been so extensively worked for this purpose.

Apart from the flint, East Anglia may be regarded as a stoneless region, and its comparative isolation hindered the importing of stone. Flint thus acquired great importance as a building stone, although bricks were very extensively made from local clays. Various stages in the use of flint may be noticed: in some of the earliest buildings untrimmed flint nodules were used with much mortar, but very remarkable workmanship is seen in many later buildings, the flint being dressed and used with very little mortar.

Around the coasts, especially in those places where many flint pebbles occur on the shore, a curious effect is made by the projecting rounded 'kidneys' of flint, the use of which is generally combined with brick, which form the corners of the cottages. This type of cottage is characteristic of many villages near Cromer, as at Trimingham, Weybourne and Kelling. But in many other parts away from the coast, even where brick is mostly used for the houses, the churches are still nearly always of flint. Among the oldest churches, round flint towers are common, the round form having perhaps been used because of the absence of any suitable

Plate 13. *The front of the old Guildhall, Norwich, showing the use of flint in buildings. A limited amount of freestone has also been used in this case for decorative effect, particularly in the 'chequerboard' work.*

large stones to form quoins or corner-stones. Many of these round towers date from the tenth or eleventh centuries; in Norfolk they are a most striking part of the landscape, 119 out of an English total of just under 180 being found in that county, with forty-one in Suffolk and eight in Essex. But the larger square towers are no less impressive or varied; that at Winterton, north of Yarmouth, not only dominates the red-roofed village but affords a view out to sea over the sandhills which hide the village.

In the building of many of the larger churches a certain amount of freestone, brought from the stone belt to the west, has been introduced, not always with happy results, for flint is a peculiar building material giving a curiously speckled effect which does not always fit in with so different a stone. The church of St Nicholas at Yarmouth is a good example in which limestone has been used for doors and windows and for places where it is desired to introduce ornament, while Cromer and Southwold churches show some remarkable flint work, as well as Norwich Guildhall (Plate 13).

This rather cold and unusual effect of the flint buildings is the most notable architectural feature in the landscape; with this is associated the lack of decoration in the smaller churches and the quite beautiful effects produced in some of the patterned buildings with their curious chequer-board design, notably in the old Guildhall at Norwich and in the church at Southwold. And added to this, the frequency of thatched roofs, as at Acle in Norfolk and Theberton in Suffolk, gives to these village churches a peculiar attractiveness.

Yet in many areas the cottages are for the most part of brick, usually of bright red colour and with red pantiles, and it is these which largely determine the colour of the villages, though in parts of Suffolk a dull yellowish-brown brick is frequent. Among these old brick buildings, more particularly in the coastal towns, are houses of distinctly foreign character, with curved gables and highly decorated circular chimneys, which are believed to owe something to the influence of the Flemings. The clays for brickmaking were dug from the superficial rocks of which such a variety is present. Here, too, are obtained the flint gravels which go to the making of the light-brown gravel lanes of the area.

Much of the east of East Anglia, from Weybourne (west of Cromer) by Norwich to Ipswich, is made up of marine and estuarine gravels and shelly sands of very variable type, known collectively as Crag (Chapter 22). They are not of great antiquity, as we have already noticed, but they were formed immediately prior to, and during the early stages of, the Ice Age. They are loose deposits, rarely of any great thickness, and were formed as sandbanks or as estuarine deposits near the mouth of a great river; at that time Britain had attained the general structure familiar to us, but was still linked to the Continent, and the North Sea was a great gulf into which the Rhine flowed and was joined on its left bank by the Thames, the deposits of these rivers contributing to the formation of the Crag.

Much of this area of crag deposits, as well as much of the Chalk itself, was further modified during the Ice Age, when ice sheets several times invaded the area. Between some of these advances of ice into East Anglia were intervals when the climate was mild and when prehistoric man lived in the area, but from the landscape point of view the most important effect was the deposition of great tracts of till, gravel, sand and loam. The details of the distribution of the successive deposits need not be discussed here. It is sufficient to notice only the most striking features, although this is classic ground for those intent on working out the details of glaciation, from such sites as West Runton cliffs in Norfolk, the old Hoxne brickyard in Suffolk and Bobbitshole near Ipswich.

At the time of maximum glaciation, the whole of East Anglia was covered by ice, and deposits from these ice sheets cover much of the area, and influence the relief and soils and hence the agriculture and character of the area. These deposits vary; they include the 'good sands' of north-west Norfolk, where on poor sheep-walks the four-course Norfolk rotation system of agriculture was developed by Townsend and Coke, and where many large farms and fields are found today, together with the stately homes built with the profits from the Agricultural Revolution. They also include the 'brickearths' of central East Anglia, which yield good loamy soils and which have through most of history been prosperous agricultural areas, today largely arable. The deposits also include sands and gravels which have profoundly modified some areas to be described separately below.

The Cromer Ridge

The most notable feature of glacial origin is the Cromer ridge, a belt of high land rising in parts to over 300 feet, which extends from near Cromer westwards to Salthouse and Holt. In detail this belt of high ground consists of a series of small ridges arranged more or less in the same direction. Along its northern border it rises sharply from the low ground near the coast, but its surface slopes gently to the south; seen from near Sheringham its wooded slope forms a very impressive feature in this area of gentle relief. To the north of the ridge are small, outlying masses of similar character, and from these the best views of the area may be obtained: the small hill about a mile west of Weybourne is an excellent example, its bracken and gorse-covered slopes contrasting sharply with the fields below. Great expanses of the main Cromer ridge are covered by heather and bracken, and indicate its sandy and gravelly nature.

Briefly, the Cromer ridge is a great moraine or series of moraines left by glaciers which advanced from the north and north-west. Its material includes much Chalk, and also rocks from Scandinavia, from Scotland and from the north of England. Generally the rocks from more distant sources are less common than those from areas over which the ice sheet had but lately passed, and therefore the rocks which occur immediately north and north-west of Cromer are more plentiful. Chalk especially is very abundant, sometimes in great masses. Near Holt and Weybourne the quantity of Chalk is so great that the material has been burnt for lime. The irregular hummocky surface represents the original inequality of the deposits, but the structure is far from simple. This can be shown from an examination of the cliffed coast between Weybourne and Happisburgh, for there glacial material derived almost entirely from early ice advances of the Ice Age is cut transversely and its nature is shown in the cliff sections. The sea is rapidly attacking any cliffs unprotected by walls or similar structures, and sections of tills, sands and gravels can be seen, with the earlier series of glacial deposits contorted also by a fresh ice advance. Here also, near West Runton, huge masses of chalk, some hundreds of yards long, have been moved by the ice and deposited some distance from their original position. At the base of the cliff the peaty deposits of the Cromer Forest Bed series can sometimes be seen. These beds are thought to have been formed during an early, warm, interglacial period in lake and estuary conditions, and yield many plant and animal remains.

Parts of the ridge form a watershed, although it is so new a feature (formed during the Ice Age). It is clear that the rivers of this area have nothing like the antiquity of those of the London basin, where, it may be remembered, the essential features of the drainage were determined a considerable time before the Ice Age. Few of the present valleys in East Anglia were cut until part of the Ice Age had passed and some till had been laid down over much of the area. Here again, therefore, the landscape of this region may be called 'new'

Breckland

While the Cromer morainic ridge is the most impressive feature resulting from the glacial advances, other wide areas in East Anglia have been profoundly modified by the glaciations. The most interesting of these is Breckland, a great stretch of largely open heathland some 400 square miles in extent, which lies around and especially west of the old town of Thetford. Here resting on the Chalk is a cover of superficial deposits, consisting mainly of gravels and sands deposited during a stage of the Ice Age. This parent material gives only thin, gritty soils mostly unsuitable for cultivation. Much of the area is covered by heather or poor grassland with belts of gnarled Scots pines. Also found are many plants which are scarcely known elsewhere in England. With the thin vegetation and porous soil the surface rapidly becomes dry and powdery and the sand is swept up by the strong winds, forming great dust storms. Thus the deposits have been redistributed to some extent by wind action. This unusual tract of steppeland was at times inhabited by early man (Grime's Graves has already been mentioned), but is now very sparsely populated, lack of water supplies being a significant feature in this respect. The occurrence of porous surface deposits over likewise porous Chalk makes for few streams and reduces the prospect of water in shallow wells. The villages are thus, as a legacy of earlier days, almost entirely confined to the valleys of the few rivers, and for miles at a time there are few dwellings even near the main roads. Thetford is the most important settlement in the area, at the confluence of the Thet and Little Ouse, where these rivers were crossed by the Icknield Way.

This great area of open land and of huge estates was formerly little modified for many centuries. The difficult soils and rabbits made cultivation unrewarding, although some estates are now productively farmed by modern techniques. Shelter belts, mostly of pine, have given some protection to the less infertile tracts. The Forestry Commission has acquired and planted large areas since the 1920s, more than 30,000 acres having been planted with conifers, although with hardwood trees as screens by the sides of the straight roads. The public is denied access to an area in the north-east of Breckland which is an Army training area. A number of shallow meres pose interesting problems as to the source of their water.

There are other expanses of heath and common near the Suffolk coast, in an area known as the Sandlings, covering much of the coastal strip between Southwold and Ipswich. Here also are patches of superficial sands and gravel, where early man has left traces, notably in the Sutton Hoo ship burial. In modern times the area is partly heath and common, partly cultivated in large fields and partly forests. These tracts are bounded sharply by the rich agricultural lands which occupy most of East Anglia, and by undulating green patches with so many belts of trees as to give the appearance of lightly wooded country.

The Broads

The Broads is one of the best known and unusual areas of East Anglia, and here recent research has brought about a series of spectacular discoveries. This is an area of flat meadows, crossed by

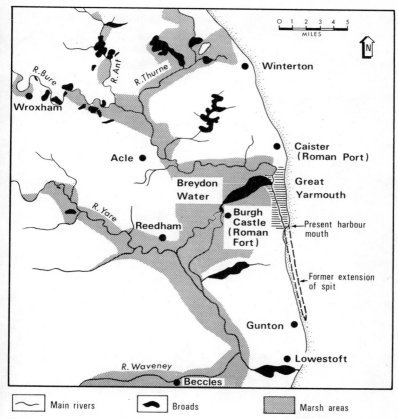

Fig. 16. *Part of Broadland, showing the former estuary and the site of Great Yarmouth.*

winding willow-bordered roads and traversed by meandering streams. These merge so imperceptibly into the wide areas of shallow water, that the traveller often gets the odd impression of white sails gliding among fields. These reed-bordered stretches of water, found especially in the valleys of the Bure, the Yare and the Waveney, probably need no description. It is easy to visualize how they were formerly thought of as naturally occurring in the process of silting up of an estuary. Today the area of open water in the

Plate 14. *The Norfolk Broads. The river Bure winds through this picture, with, in the foreground, Hoveton Great Broad beyond the river and Salhouse Broad nearer to the camera. In the right background are parts of Hoveton Little Broad and Decoy Broad. The photograph shows the general separation of river and broad, except through narrow connecting channels. The encroachment of sedge fen into the area of open water is also prominent. The land beyond the broads is of Norwich Crag, largely covered with loamy glacial deposits.*

Broads, except where rigidly controlled, is being rapidly diminished by the encroachment of the reeds.

Certainly the Broadland area must have been, probably as late as Roman and Anglo-Saxon times, a wide open estuary, with exits to the sea near Horsey, about twelve miles north of Yarmouth, at Yarmouth itself and at Lowestoft (Fig. 16). Silt deposition must have taken place in the quieter waters inside the estuary, more particularly at the seaward end, with Breydon Water as a present reminder of these conditions. This would have given rise to brackish, swampy conditions at the upstream end, where the Broads are found today. Here there would have been dense vegetation, perhaps alder thicket and sedge fen in the main. Death and regeneration of this vegetation led to the formation of thick layers of peat.

It is now thought that the Broads themselves are vast, flooded, medieval peat diggings, the peat having accumulated during swampy conditions, and been dug subsequently for fuel. The digging was made possible by a relative rise of the land during medieval times. The main evidence for this conclusion came from a painstaking investigation by boring of the nature of the beds of the Broads. It is well known that the Broads are not widenings of the main channel of the rivers, but are connected with the rivers only by narrow side channels (Plate 14). Baulks and strips of peat were found remaining in the beds of the Broads, notably along what are now parish boundaries, indicating removal of peat by man. At the same time evidence was found, from excavations for the construction of a power station at Great Yarmouth, that sea-level was relatively lower than at present from the eleventh to the thirteenth century, after which flooding began. Documentary evidence was found in support of these conclusions, that peat digging took place from about the eleventh to the thirteenth century, after which the areas stripped of peat were gradually flooded and formed the Broads as we know them today.

The Coastal Scenery

The coast of East Anglia has perhaps more features of interest than the inland areas. The interest of the cliff sections of Norfolk in showing the glacial deposits has already been mentioned, as have the relative changes of land- and sea-level which allowed the digging of the Broads. There is an immense variety in the East Anglian coast. Along the north coast of Norfolk, west of Weybourne, salt marshes front an ancient cliff line, the marshes being in their turn protected by spits and barriers of sand and shingle, notably Blakeney Point and Scolt Head Island. North-east Norfolk meets the sea along the twenty-one miles of eroding cliffs from Weybourne to Happisburgh: these cliffs are succeeded in their turn by a single thin line of dunes which denies the sea access to much low-lying marshland in Broadland. Dunes and cliffs alternate from Winterton past Yarmouth and Lowestoft to Aldeburgh, where the magnificent shingle spit of Orford Ness is joined to the land and whence it deflects the river Alde southwards for about nine miles. To the south again, low cliffs and marshes alternate with estuaries where sea water penetrates to such towns as Woodbridge, Ipswich, Manningtree, Colchester and Maldon, providing port facilities, and pleasure for the yachtsman.

The coast, as well as the Broads, has been greatly affected by relative changes of land- and sea-level. Sea-level has been rising, with only minor oscillations such as that which made possible the peat digging in the Broads to break the trend, since the last retreat of the ice from Britain. When very large amounts of the earth's total water supply are frozen as ice, then much of this water must come from the oceans and sea-level is lower. There is a slight compensating effect in that the great weight of ice or of new sediment on the land areas tends to depress the earth's surface in these areas and when the ice melts the land will tend to 'rebound' to regain its former position. It is the complex interaction of these two factors which determines the relative level of land and sea at a particular place. The results in East Anglia have been a rapid rise in sea-level with the retreat of the ice, reaching present sea-level for the first time several thousand years ago, with oscillations ever since. East Anglia is subsiding relative to the sea at present, and it appears that this trend has been steady since the thirteenth century.

This relative rise in sea-level has drowned the estuaries of south Suffolk and Essex, while farther north estuaries, notably the Broadland estuary, have silted up and peat has formed, and spits and barriers have been formed across the mouths, thus smoothing the coastal outline.

The Broadland estuary outline has been smoothed by the growth of a spit or barrier across the mouth at Yarmouth, on which that town is now built. In Roman times there was a port at Caister-on-Sea, and a Saxon Shore Fort at Burgh Castle, then probably on a peninsula commanding the tidal estuary, now overlooking the narrow Waveney channel near the entrance to Breydon Water (Fig. 16). Yarmouth itself is built on the narrow spit between the river and the sea and this has controlled the shape of the town. The sea front and resort extends along the coastal face of the spit, the long main streets broaden into the market-place, and formerly the narrow 'rows' led down towards the riverside. Modern industrial development is taking place at the southern end of the spit. In the fourteenth century, however, the Yarmouth spit was much longer, extending then for about eight miles to near Gunton. Because of its interference with navigation it was cut through at various dates, the present mouth of the Yare being fixed in the sixteenth century and artificially maintained ever since. A slightly similar situation exists at Lowestoft with the Denes as a former low foreland, and Oulton Broad and Lake Lothing ponded back, but artificially connected with the sea.

When this Broadland estuary was an open tidal estuary, there were two islands to the north and south of Yarmouth – Flegg and Lothingland. Flegg particularly had many village names of Scandinavian origin, ending in 'by', for example Hemsby, Scratby, indicating Danish settlement or conquest in Anglo-Saxon times.

The growth of these spits and barriers implies the supply of great quantities of beach material and we should inquire as to the source of this material. There is fine material and sand forming alluvium in estuaries and marshes, and on beaches facing the sea there is mainly

sand and flint shingle, the latter derived originally from flints in
Chalk. It was thought until recently that all this material was derived from the rapid cliff erosion, but this is now being questioned.
It is now thought that some of the material, particularly flint
shingle, has been swept up from the sea-bed during the post-glacial
rise in sea-level as the last ice sheets of the Ice Age melted. This
process has already been described in considering Chesil Beach in
an earlier chapter (p. 124). If this is true, then suitable material on
the sea-bed is or was plentiful, since the whole of the North Sea off
this coast was at one time glacial. These shingle beaches in East
Anglia are also 'fossil' beaches, with no major source of replenish-
ment at present. In south Suffolk and Essex it seems that little
material was available from these sources, so there are no shingle
barriers, and the estuaries are wide inlets.

The movement of beach material is also therefore of some interest. Pebbles are moved only by waves, whereas sand and finer
material can be put into suspension by waves and held in suspension by the turbulent movement of the water while they are being
transported by the currents. At least from Sheringham eastwards
and southwards the net movement of material along these coasts is
in this easterly and southerly direction. Evidence for this can be
seen in the way in which material is piled against the west or north
side of the many groynes along the coast. It was formerly thought
that from about Sheringham westwards, material moved in the
opposite direction, since Blakeney Point and Scolt Head Island are
growing westwards, but it is now thought by some that these formations are built mainly of sand brought from the west by wind,
waves and currents, and shaped by wave action, while the shingle
forms a kind of protective covering on top of the sand, moving to
and fro with the waves. Salt marsh and dunes grow inside the outer
beaches of these formations, providing a suitable habitat for many
migratory and resident birds. Orford Ness in Suffolk is another very
fine shingle spit, deflecting the river Alde southwards about nine
miles.

Another source of material is from cliff erosion, yielding largely
sand and fine material, with few pebbles. The cliffs are being rapidly
eroded here, losses averaging in places three feet or more a year,

since the glacial deposits do not resist erosion as does hard rock. Many of the cliffs are nearly vertical, waves attacking the base and removing the material, but on some lengths, especially between Cromer and Mundesley, the presence of ground water in the cliffs causes the collapse of large areas at a time, with tongues of material spewing on to the beach below. This debris is then gradually removed by wave action and the slope steepened for the process to be repeated.

·Some of the sand derived from these and other sources goes to form the sandy 'nesses' which occur along this coast, for example Winterton Ness, Benacre Ness, Thorpe Ness, the North Denes at Yarmouth, and in earlier days probably Lowestoft Ness. It is thought that these are formed largely by tidal current action, in which the ebb and flood tide currents form separate channels with banks thrown up at the margins. If one of these margins is the shore, then a sandy ness will result. That at Winterton fronts an old cliff line that was almost certainly washed by the sea at least intermittently as late as the eighteenth century, pointing to the fickle changes from erosion to deposition which can occur along this coast, possibly due to changes in the form of the offshore banks. Dunwich is another example, the site of an early see and some twelve churches. Most of the site has been lost to the sea, but the cliff now appears to be stable.

It is probable that a number of settlements have been completely lost in East Anglia, as in Holderness (pp. 74–5), by the landward movement of cliffs and dunes. One of the most spectacular instances of such destruction is on the Norfolk coast where the church of Eccles north of Yarmouth was gradually submerged by moving sand-dunes. Lyell called attention to this church in his *Principles of Geology*; the tower was of a familiar Norfolk pattern, round in the lower part, octagonal above, and although little information exists regarding the date when it was built it is unlikely that the top was added before the sixteenth century. It is thus highly probable that there was no apparent danger from the sea at that time, but by the beginning of the next century the inhabitants were petitioning for a reduction of taxation on the ground that much land had been destroyed by the sea. Sand-dunes ultimately covered the area formerly occupied by part of the village, and by 1839 they had advanced so far as partly to bury the tower, while before 1862 they had been carried inland by the wind to the land-ward side of the tower, leaving it exposed to wave action. The tower collapsed in a storm in 1895, but the foundations can still be seen on the beach after a scouring tide.

Walls and groynes have been built along the coast to try to check this movement and destruction, and resorts such as Sheringham and Cromer are assuming a salient position, holding back beach material by groynes as best they can. Destruction still takes place: north of Yarmouth the dunes were breached and flooding resulted, at Horsey in 1938 and Palling in 1953. Walls have been built to check a recurrence of these disasters. It is important to speculate on what will happen if the whole coast is eventually encased in concrete with groynes, and no further supply of material is available from coastal erosion.

9. The Fenlands

If the sea-level were now to be raised by about twenty feet in the vicinity of the Wash, that bay would be extended almost to Skegness, Sleaford, Peterborough, Cambridge, and Brandon in Suffolk (Fig. 17). A large number of islands would be seen, and on these would survive many of the existing settlements of the area, perhaps the best example being Ely, with its cathedral visible for many miles. This enlarged bay would define the Fenlands, wide areas of flat alluvial land, rare in British topography, forming level expanses with few breaks for many miles. Now some of the best agricultural land in England, the area has formerly been in turn an arm of the sea, woodland and nearly impassable swamp.

Four main rivers drain the Fens, the Witham, Welland, Nene and Ouse, and they all drain into the Wash. The Wash is a major break in the Chalk, between Hunstanton and Skegness, presumably carved by these four rivers when the sea was at a lower level, and the rivers were following the lines of what are now the deeps of the Wash. Inland of the Chalk was a wide expanse of Jurassic rocks, mainly clays (the Oxford, Ampthill and Kimmeridge Clays) offering little resistance to erosion, and thus a broad, shallow depression was formed. During the Ice Age, at one period the Fens were covered by ice, some tills and gravels were deposited, and at a late stage the area was a lake, ponded up by ice in the Wash, and possibly draining along the line of the present courses of the Little Ouse and Waveney rivers to Lowestoft and beyond (Fig. 17). During the post-glacial changes in sea-level the Fens were at some time a shallow arm of the sea, or swamp, with silt being deposited at the seaward end, probably aided by the roots of salt-marsh vegetation as a silt trap. Meanwhile farther inland, brackish or fresh

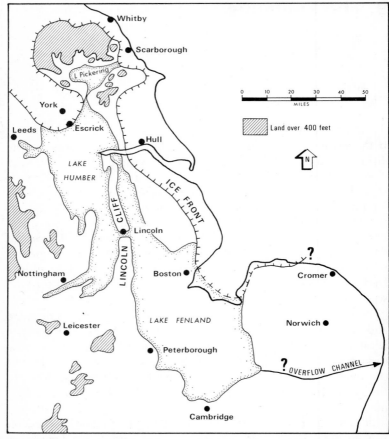

Fig. 17. *The probable limits of former glacial lakes in eastern England. These lakes existed during a late stage of the Ice Age, with the ice front in approximately the position shown. It is by no means certain that all these lakes existed simultaneously, so the diagram may represent conditions over a long period. North Yorkshire is shown in more detail in Figure 4.*

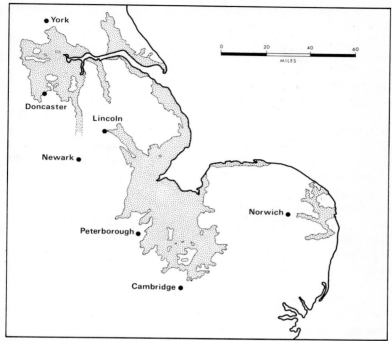

Fig. 18. *Alluvial areas of eastern England (after J. A. Steers). This map should be compared with Figures 16 and 17.*

water conditions prevailed with dense vegetation in the swamp, and peat forming from its decaying remains. Slight changes of level would allow the silt to encroach temporarily on the peat and vice versa, and together with slight changes of climate would have allowed a complicated alternation of woodland and swamp. Further complications result from possible subsidence within the bay, and thus, with changes in relative sea-level, buried forests are now found at depths below present tide levels.

Essentially this gives the simple divisions of the Fens into the seaward silt fens and the inland peat fens, with 'islands' of the older rocks, as at Ely or Boston, or glacial deposits, as at March, and gravel or sand shore deposits around the margin.

Conditions during Roman times were such that they were able to settle the silt fens and make use of the salt marshes, but conditions deteriorated during Anglo-Saxon times, and the Fens were largely abandoned, becoming a frontier region between East Anglia and Mercia, contributing to the former's isolation. This isolation is demonstrated dramatically by the Devil's Dyke near Newmarket race course, a boundary ditch and bank of this period, some seven miles long, lying athwart the narrow chalk corridor between impassable Fen to the west and boulder-clay-covered, wooded upland to the east.

This change to swampy conditions is thought to have come about as a result of a slight change in sea-level and possibly the disrepair of Roman artificial drainage. The conditions survived for centuries; Domesday confirms that the silt fens and fen islands were inhabited, while the peat fens were not settled. Be they legend or truth, the tales of Hereward the Wake indicate the conditions of his time. By about the thirteenth century at least the silt-fen area of Marshland in Norfolk must have been prosperous, judging by the magnificent churches of the period to be found there, although the peat fens were inhabited only by a few fowlers, fishermen and sedge-cutters. A sample of this peat-fen landscape has been preserved at Wicken Fen, about twelve miles from Cambridge, where rushes, sedges and alder thickets border the shallow waters, but artificial maintenance is now required to keep the area in what are thought to have been natural conditions in earlier times.

The present condition of the Fens owes most to the hand of man. Early attempts were made to reclaim the silt fen; and although the Roman Bank is almost certainly not Roman, it is probably pre-Conquest in age. Other banks have been built to seaward of this to reclaim more land from the Wash, and the process is continuing up to the present. The successful reclamation of the peat fen was begun in the seventeenth century by Cornelius Vermuyden and other Dutch engineers, financed by among others the Duke of Bedford (hence the Bedford River and the Bedford Levels), not without resistance from the watermen, but the richness of the new land which was made available guaranteed the continuation of the efforts, giving land which was first used as rich sheep pasture before passing to arable use for high-value crops such as bulbs, fruit, vegetables and potatoes. At first windmills were used, but drainage brought further problems. The peat shrank as it dried, and wind and bacterial action also wastes the peat. No such effects apart from wind action reduced the silt fen to seaward, and so with this shrinkage of the peat, drainage became more difficult. Some open water, notably Soham and Whittlesea Meres, survived until the nineteenth century. The Holme Fen Post near Whittlesea Mere indicates peat shrinkage and loss of about thirteen feet since its drainage last century. Loss of topsoil in dust storms is now also a serious problem.

Shrinkage of the peat made drainage by windmills more and more difficult, and the lift for the water was always increasing. Fortunately steam pumps were brought to the rescue, and these have in turn been superseded by diesel and electric pumps. The water was pumped into many new artificial channels as well as the old natural channels. Most prominent of these artificial channels are the parallel Old and New Bedford Rivers, with the Washlands between them used only for pasture and temporary storage of flood water. Most of the channels now stand well above the surrounding peat level, due to shrinkage of the peat and embanking of the channels. Many of the courses of former small wandering streams in the peat are now marked by 'roddons' or raised banks, formed by silt which once filled these channels; while the peaty deposits have shrunk the silt has remained practically unchanged in bulk, so that the river course, formerly below the level of the peat, now stands out above it. The roddons have frequently been selected for farm building sites, and they stand out clearly as a change of tone in air photographs.

While the natural drainage has undergone great man-made modifications in recent centuries, these natural channels have themselves changed in the past. Until late in the thirteenth century the Nene, Great Ouse, Cam, Lark, Little Ouse and Wissey all reached the sea through Wisbech. A catastrophic storm resulted in the diversion of much of this flow to the Lynn outlet, formerly serving only for the Nar and the Gay, and the present situation has been brought about and maintained largely by engineering artifice. The problems of fen drainage are still acute, as flood water from the surrounding upland would naturally drain through the

Plate 15. *Fenland near Southery, Norfolk. In the foreground is the confluence of the Great Ouse and Little Ouse rivers at Brandon Creek. Note the raised river banks, and the roads following these. Many of the buildings are also on or near the river banks. In some of the fields the tone differences show evidence of former natural drainage channels. The village of Southery in the middle distance is sited on a low 'island' of Kimmeridge Clay, while in the background is higher ground on Gault, Greensand and Chalk with drift cover.*

Fens. and occasional high sea-levels in the Wash brought about by north-westerly winds in the North Sea may prevent the discharge of flood water for several hours around each high tide. The last disastrous floods in the area were in 1947. mainly from land water. and in 1953 due to high sea-levels. Since then a diversion channel has been built along the fen margins to take some of the flood water from the Norfolk streams to the Lynn outlet without it passing through the peat fens.

Low ground extends all along the coast of Lincolnshire. It forms a belt stretching to the coast from the old cliff bounding the chalk Wolds. Within this belt two types of scenery may be recognized. Irregular, hummocky land formed by the local till is found fringing the Wolds, while nearer the sea is almost flat marshland. The featureless surface of the marsh is almost treeless, as contrasted with the well-wooded till, but numerous islands of till stand up above the marsh, which thus repeats on a small scale the conditions of the Fens.

Some fen country is also found around the Humber river system, with a history and appearance similar to that of the main Fen areas (Figs. 17 and 18). These will be briefly mentioned in the next chapter.

10. The Heart of England

The great triangular area of low country which forms the centre of England is bounded on the west by the north–south line which marks the Welsh borders and on the east by the northern extensions of the Cotswold belt; the most southerly part of the plain tapers along the Severn valley to Gloucester. From this broadly triangular tract there are northerly projections stretching along both east and west sides of the Pennines.

While most of this region is of low elevation it nevertheless shows much variety in relief. Its hills rarely exceed 500 feet in height, but they are of several different types, and the area as a whole is very diversified. Within it there are places of unspoiled loveliness, the woodlands of Sherwood Forest and the soft green lands of the Vale of Gloucester, contrasting with industrial areas of unmatched ugliness where not an acre of the original surface remains unscarred by diggings or by refuse. The Midlands cannot be summed up in a word; one speaks of 'the Midlands, that are sodden and unkind', another of 'these most beloved of English lands', and both perhaps are right.

The Vales of Gloucester and Berkeley

This tract is simplest in its southern part, in the Severn valley north of Gloucester. Here on the west the Malvern Hills form a cliff-like boundary rising impressively from the vale, while on the east the more irregular but no less conspicuous scarp of the Cotswolds makes a definite boundary (Fig. 2). The intervening area is almost uniformly low, but through it runs a somewhat indefinite wooded ridge rising in places about 200 feet above the level of the valley. For the most part this higher ground lies over on the east bank of the river, but it is so weak a feature that the Severn several times crosses it south of Tewkesbury.

Seen from the hills on either side, the Vale of Gloucester presents a wide expanse of fertile lowland, with flat, riverside meadows situated on the level tracts of alluvium laid down by the river, and liable to frequent floods, for the Severn carries flood water from the upland areas of Central Wales. Stretching away from the river for miles, the fields, bounded by high hedgerows, still retain their rich green. From many directions the vale appears to be well wooded, but this results mainly from the numerous tall trees which form part of the hedges and border the lanes. elm and oak being most common. despite the Dutch Elm disease of the 1970s. Orchards are also common, especially in the west. A tongue of similar land is found downstream of Gloucester, between the Cotswolds and the Severn, known as the Vale of Berkeley.

The southern part of the vale is dominated by the great tower of Gloucester Cathedral situated near the centre of the old but rapidly expanding town, and just above the alluvial tract of the Severn. Tewkesbury, with its solid Norman abbey, is similarly placed some ten miles upstream, and retains more of its original character, with its streets of Elizabethan houses, half-timbered and gabled, and narrow lanes affording glimpses of river and meadows from the main street. Whatever the charms of Cheltenham, it has much that is alien to the vale, for it was little more than a village until the early eighteenth century, when its waters were accidentally discovered, and its subsequent progress has been that of a successful spa. The attractive Regency architecture and wide boulevards of Cheltenham reflect the date of its development, for it grew up late enough to be influenced by formal town planning. In this respect it contrasts with the apparently haphazard but nevertheless picturesque street patterns of the older towns farther north, such as Tewkesbury, Pershore and Evesham, which are characterized by their half-timbered buildings.

Many of the villages are scattered and irregular in form, sometimes straggling along the several roads leading from the centre. As in the old towns, so to an even greater extent in the villages, stone is rarely used in building, and cottages of brick or half-timber, thatched or tiled, are characteristic of the older houses of the whole area, though stone from farther afield (and often from the Cotswolds) has been used in building the churches. It is at once obvious that this is not a stone district, almost the only stone being in the villages along the low ridge referred to above. Apart from this, the whole area is underlain by marls and clays, which normally give rise to a heavy soil. This clay belt was originally densely wooded, and it was only gradually that parts of it were cleared for cultivation, long after the tracts of lighter soil (for instance on the Cotswolds) had been developed. On such clay ground the dominant natural vegetation was probably oak woodland, and the dense forests together with the marshy nature of this low, clayey country rendered passage through the area difficult. It is probable that, in such country, clearances were made at a number of spots, many of which later became the sites of settlements. Thus numerous small villages are now found irregularly scattered at intervals of only a mile or two through the area, which has never developed the large, concentrated villages to be found in the areas of lighter soil. Moreover, in the clay areas water supplies, at least of surface water suitable for early settlements, were so frequent as to cause no difficulty.

The clays were dug for brickmaking, in the absence of suitable stone, and small brickworks formerly existed in very many of the villages in this belt, supplying also tiles and, later, drain-pipes for agricultural land. The clays are of two types: in the west of the area they are deep red in colour, and give rise to fields of characteristic appearance, while east of the Severn and extending to the slopes of the Cotswolds they are generally blue or grey-blue, turning to a dirty yellow at the surface. The latter clays give heavier and stickier soils than the red marls of the west, but when used for brickmaking both yield bricks of a red colour. The difference in colour of the clays is due to a difference in the iron compounds which are present, and the baking of the blue clay brings about a change which turns the bluish colouring-matter to red.

The clays or marls of red colour belong to the rocks known as Keuper Marl, a part of the group named the New Red Sandstone: the blue clay represents the lowest part of the Lias, the bottom group of the Jurassic. Both groups are dipping to the east and pass under rocks which form the scarp of the Cotswolds. The base of the Lias includes some beds of blue limestone (the blue Lias) which give rise to the low ridge mentioned above. On the western edge sandstones are present amongst the red rocks, and as these rise on to the shoulders of the Malverns they form a small area of lighter soils and rougher country in the area around Bromsberrow Heath.

It must not be thought that the course of the Severn through this tract of green lowland has been determined by the presence of higher ground to west and east, for in all probability the vale was not a marked depression when the Severn started on its work of erosion. In other words, the vale has been carved out by the river and by weathering agents; it is only when this conception of its origin is grasped that the real significance of river work can be appreciated. Standing either on the Malverns or on some viewpoint on the Cotswold edge, such as Leckhampton Hill, the immensity of the valley is so impressive that it may be hard to realize that the rivers almost unaided have been able to carry away the vast amount of material which formerly filled it. Such a contemplation of a great valley affords a demonstration of the length of geological time, for the carving of this valley (although it has taken some millions of years) is but a recent feature of the landscape when viewed in relation to the time during which the rocks underlying it were formed.

An important factor in the later history of the Severn valley evolution was the presence of large ice sheets which originated from the north and west. Thus the immense volumes of meltwater, although not playing a significant part in valley deepening, have nevertheless contributed a great deal of aggradational material in the form of alluvium and river gravel into which the river has entrenched itself periodically as the sea-level fluctuated during the Ice Age. The terraces can be traced upstream from the Vale of Gloucester into the middle Severn and the Warwickshire Avon river systems and have played no small part in the layout of the riparian settlements. The presence of fossil peats and other organic remains in association with some of the lower river terraces has enabled a fairly accurate assessment to be made of their age, all of them having been formed within the last 50,000 years.

The Midland Plain

Many students of river development have thought that when the English rivers began their work of carving the present landscape, not only was there no Vale of Gloucester, no Vale of Evesham and no Midland plain, but the drainage from parts of Wales (including the Wye, Severn and Dee) flowed eastwards or south-eastwards down a gently inclined surface which planed both older and newer rocks, long before the present relief of the English Midlands and the scarplands had been etched out by subsequent river incision. As the landscape was lowered the original inclined plane was dissected by such tributary streams as the lower Severn which was running at right angles to the direction of the older streams and was able to capture many of their headwaters. The rapidity of this river capture may have been assisted by the sinking of the Bristol Channel due to faulting, a hypothesis which is referred to in Chapter 19. It has been suggested that the original inclined plane was an eastward extension of the Low Peneplain of Wales (which is discussed in more detail in that section) and that the last surviving remnants of this surface can be seen in the highest summits of lowland England. A glance at a relief map will show a remarkable coincidence of highest summit elevations in the Midlands, despite their geological variety. Thus the Clent Hills (New Red Sandstone breccias) reach 1,035 feet; the Lickey Hills (Pre-Cambrian to Lower Silurian), 956 feet; Cannock Chase (New Red Sandstone), over 800 feet; Charnwood Forest (Pre-Cambrian), 912 feet; Cleeve Cloud (oolitic limestone), 1,083 feet. It is tempting to regard these culminating heights of Midland England as survivals of the surface from which all the present landforms have been carved, largely by river activity, and as a continuation of the late Tertiary summit surface already noted on the chalklands. In this way the Midland region has become pre-eminently the land of the great English rivers, for apart from the Thames, all the larger rivers flow across it for considerable parts of their courses.

For ease of description it is proposed to divide the Midlands into two, a division which reflects the differences in geology. In the northern part the presence of Coal Measures with their associated iron ores and fire clays has led to large tracts of industrialization and great urban expansion: the so-called 'Black Country', the Birmingham conurbation, Derby, Nottingham and Stoke-on-Trent are typical of this section of the Midlands. In contrast, the southern part of the region, lying between the Midland coalfields and the Cotswold scarp, is almost entirely devoid of industrial landscapes, and in Worcestershire and Warwickshire the rural landscape of the plain stretches some forty miles from west to east. Here is the red English plain in its finest development, extending along the valleys of the Severn and the Warwickshire Avon, the England of Shakespeare's country, with its roads fringed with great trees, with its red ploughed fields making the green of the hedges seem a deeper tint, with innumerable villages of red brick or black and white thatched cottages set along twisting roads, the narrow gardens fronting the road bright with common flowers. Where nearly every village is 'typical' of the area, it is useless to name examples, but none is more charming than Harvington, where every turn in the road brings a new picture into view.

The market towns of this area are nearly all delightful; Warwick, Evesham, Pershore and Worcester all have their half-timbered houses, and when stone comes into use, it is frequently, as in Warwick Castle, from some of the red or grey sandstones which occur below the red marls. In spite of its temptations, Stratford-on-Avon has so far contrived to keep something of its own quality, and few of these old towns have been touched by large-scale industries, for there are no coalfields in the southern tract. Along the Severn itself, no coal-bearing rocks are exposed until the Forest of Wyre coalfield is reached near Bewdley, where, as at Coalbrookdale, the navigable river led to early exploitation of coal and iron deposits. The amount of material available was not comparable with that in the larger coalfields, however, and this trade gave rise to no important industrial developments in the lower Severn valley.

The Avon, the largest tributary of the lower Severn, and the best-known river of the southern Midlands, has an interesting geological history which is worthy of our attention. It appears probable that the earliest drainage of the south Midlands was north-eastwards towards the Trent river system. It has been shown by Professor

F. W. Shotton, however, that an early ice advance into the area impounded a large pro-glacial lake, termed Lake Harrison, against the Cotswold scarp. It was, at its maximum size, fifty-six miles long between Stratford and Leicester and reached a height of over 400 feet before overflows occurred into the upper Thames system by means of the Cotswold gaps at Fenny Compton and Dassett (Fig. 19). Thus the Avon is a very recent river, having been initiated by the meltwater streams from the retreating eastern ice lobe.

Turning to the more northerly part of the plain, we find that wide areas have been industrialized, and the whole region has changed its character during the last 180 years. Few of the older towns remain unaffected, and most of them are completely changed by the products of this later period. In parts of the north Midlands it is impossible to imagine the beauty that once existed, but the ugliness which has replaced it is just as surely a result of the geological structure, for it is the distribution of materials which has determined the areas which have become rich and populous. Here in the north Midlands are the coalfields of north and south Staffordshire, Warwickshire, Leicestershire, Derbyshire and Nottinghamshire, too varied in activity and scenery to be described as a whole.

It is a wide area chiefly based on sandstones and clays. As compared with the Vale of Gloucester, sandstones are more important, but the clays and marls continue to be important, particularly since the greater rivers favour them. So the sandy areas are rather more elevated with drier soils and different vegetation, as at Cannock Chase with its extensive heathlands and coniferous plantations. Perhaps the most suitable point at which to commence the examination of this region is the north-east corner, in the neighbourhood of Nottingham and Leicester, where the Trent valley makes its sweeping curve from Burton past Newark to the Humber, keeping to the outcrop of the red marls.

One of the most characteristic rocks in this area near Nottingham is the Bunter Sandstone, on which the city was founded. It forms the steep crag on which the castle stands, and extends thence northwards through Bulwell Forest to the area east of Mansfield and Worksop. The Bunter yields a light, thin soil and although many areas are cultivated there are still great expanses of heathland with

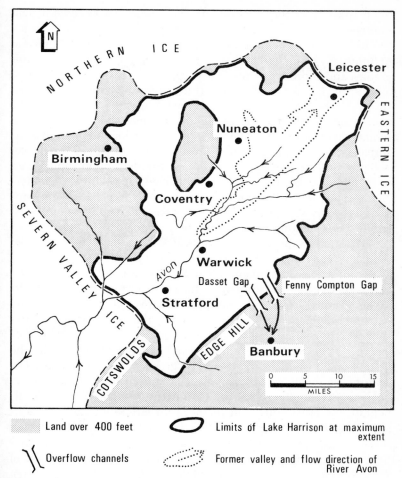

Fig. 19. *The location of pro-glacial Lake Harrison.*

gorse, ling and broom, while in the north Sherwood Forest with its oaks and dense bracken remains a vast expanse of primitive wood-land, left unworked because of the poor soil; its former extensions southwards to the borders of the city, however, have slowly yielded to the advance of cultivation. In the north, these poor soils were often utilized in the creation of large private parks, of sufficient number to inspire the name of the 'Dukeries'.

This Bunter area consists of rolling hills about 400 feet high and of unusually rounded shape, rarely showing simple scarp topography and more frequently yielding convex than concave slopes, a result of the uniformity of the thick, soft sandstones. In this they resemble the Chalk, and also in their porosity, for most of the valleys on the Bunter Sandstone are streamless, practically all the rainfall sinking into the ground. The water-table rarely reaches the surface; in many places it is at great depth, and within the Bunter outcrop very deep wells are necessary if underground supplies are to be obtained. Such wells afford water nowadays for large undertakings, but the difficulties of water supply precluded any considerable early settlement in the area except along the rare streams. It has therefore long remained sparsely populated. In the valley of the Maun are Slipstone, Ollerton and Edwinstowe, but elsewhere there are scarcely any old villages.

Recently, however, there has been a tendency for new mining towns to spread on to this part of the red sandstone country, resulting in a rural landscape interspersed with collieries and mining villages, graphically described in some of the novels of D. H. Lawrence. Early coal-mining was mainly confined to places where coal emerges at the surface or where it can be mined at little depth. The belt of small coalfield towns which grew, particularly in the last century, on the uneven country extending from Ilkeston to Chesterfield, marks the outcrop of the Coal Measures, the 'exposed' coalfield of east Derbyshire. But the need for large supplies has gradually led to coal-mining being carried farther east, under the newer rocks, where the Coal Measures can be reached only at greater depths: recent development has thus tended to the exploitation of the 'concealed' coalfield. But while the former pits were often small, the deeper and more scientifically planned pits can only be economically worked if they deal with large areas underground, so that the collieries which are being sunk in the rural areas are fortunately few in number, though they carry with them on to these areas vast new populations. It was during the drilling of test boreholes to ascertain the structure of the buried coalfield that an

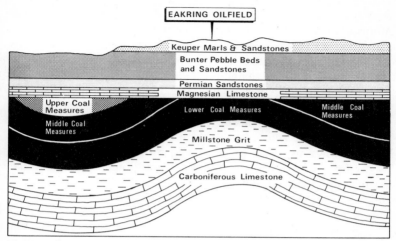

Fig. 20. *The structure of the Nottingham oilfield. The oil accumulates in the upfold of the Millstone Grit.*

interesting discovery was made near the little village of Eakring. After boring down through 2,000 feet of New Red Sandstone (Triassic and Permian) and Lower Coal Measures crude oil was encountered in the Millstone Grit without having passed through the Upper and Middle Coal Measures which were expected to be present. Figure 20 illustrates the structure that was discovered beneath the unconformable cover rocks of New Red Sandstone. It will be seen how the Carboniferous rocks have been folded and partly worn down before the deposition of the newer rocks. The eroded anticline, where the Coal Measures have been destroyed, is the structure in which oil has accumulated within the Millstone Grit and this has led to extensive drilling operations in the area between Worksop, Southwell and East Retford. As a result, dotted amongst the fields and woodlands of central Nottinghamshire are features alien to the English landscape, oil-drilling derricks and pumps, which produce a small but useful annual output from Britain's only significant mainland oilfield.

Mostly the Bunter Sandstones are soft and are very easily excavated: in the Nottingham district are great numbers of caves which have been dug out, and which generally remain dry owing to the porosity of the sandstone and the fact that they are situated far above the water-table, so that any water rapidly percolates down-wards. These excavations have been used as dwellings, and the famous 'Trip to Jerusalem' inn, beneath Nottingham Castle, ex-tends into the rock itself, while the dungeons and 'Mortimer's Hole' are also hewn in the solid rock beneath the castle.

A few miles to the north of Nottingham, on a hillside near Bramcote, stands the remarkable pillar known as the Hemlock Stone, always an object of interest to those who live in the city. Its occurrence has been attributed to Druids, and of course to the Devil, who is believed to have thrown it at a good Abbot of Lenton, and to have fallen a few miles short in his aim. But whatever the stories may say, this stone was not transported here either by Devil or by ice (which has done much work in dropping huge blocks in unexpected places), for it belongs where it is found, and is effec-tively rooted on the hill. It stands out, as on a larger scale do many mountains, because the material around it has been worn away: it is, in short, like a small outlier, which has resisted erosion be-cause it is a little harder than the rock round about. In this case the hardness is due to the fact that the sand grains are held together by a cement of barium sulphate, small quantities of which are irregu-larly distributed in the Bunter Sandstones. This insoluble cement-ing material has enabled this mass of rock, as well as the larger cappings of neighbouring hills, to resist disintegration.

The Keuper Sandstones give rise to quite a different type of scenery from that of the Bunter, for they mostly occur in thin beds among the marls, especially near the base of the group, and form sharp escarpments with steep or alternately convex and concave faces quite unlike the generally convex hills of the more uniform Bunter. The high ground to which they give rise runs north-east-wards from Mapperley, and on account of the dip the scarps fall away sharply to the west, and more gently to the east into the Trent valley.

These sandstones thus occupy an interesting position between the Bunter Sandstones on the one hand, with practically no surface water or shallow wells, and the red marls on the other which are so generally impermeable that almost all the rainfall flows away in surface streams, and scarcely any water at all is available from shallow wells. In this intervening tract there are numerous springs and shallow wells, and this group of rocks is known locally as the 'Waterstones'. It is not surprising, therefore, that along this outcrop and just below it are many villages, Arnold, Calverton, Woodborough, while others are situated along the steep-sided little valleys or 'dumbles' which are cut into the surface, as at Gedling, Lambley and Lowdham. Farther north, the old towns of Southwell and East Retford are situated in this same belt.

A far wider area is occupied by the outcrop of the red marls, which form a belt ten miles in width. East of the Trent this is occupied by low country, ill-drained and peaty in some parts, as near Gotham, broken by small ridges where sandstone bands produce minor scarps. This wide clay plain has much in it that recalls the western part of the Vale of Gloucester, and its arable fields show the same deep red. But although the villages are fairly numerous they are not always so picturesque in their setting as those of Gloucestershire, nor are the cottages individually so attractive; they are almost universally built of red brick made from the clays, rarely thatched and usually tiled. Timbered cottages are not so common as in the south and west, but in strong sunlight or at sunset the rich glow of the red cottages in such villages as Orston and Cotgrave is very pleasing. Generally the churches are of stone, grey or brown or yellow according to its derivation, and tall spires are characteristic, but brick has been used in church-building in a few cases, the diapered tower of Edwalton being a noteworthy example.

Across this area to the east there rises the low but distinct wooded ridge formed by the limestones of the 'Blue Lias', running irregularly across the low ground, in the well-known Bunny Hill and passing through Elton and Kilvington to beyond Newark. East of it the Lias clays form even ground scarcely broken by any feature until the Lincoln Cliff, or farther south, the boundary of the Vale of Belvoir, is reached.

Before leaving the description of the rocks of this part of the Midland plain it may be useful to tabulate those which are represented, so that their relative positions may be clear. In the table the oldest rocks are placed at the bottom, and it will be recalled that these are missing in the Gloucester area.

Lias	{	Lias clays (with Marlstone forming escarpment)	
		Blue Lias (Limestone forming small ridge)	
		Rhaetic. White 'Lias' (Limestone)	
New Red Sandstone	Trias {	Keuper {	Keuper Marls
			Keuper Sandstones
		Bunter Sandstones	
		Bunter Pebble Beds	
	Permian {	Magnesian Limestone (forming an escarpment which increases in importance in Yorkshire)	

As we have already pointed out, many of the New Red Sandstones are deep red in colour, and the others are generally but not always reddish. They form a great series, some thousands of feet in total thickness. As a whole the marls actually occupy a much greater area at the surface than the sandstones: this is particularly true, as we have seen, in the lower part of the Severn valley, but it is also generally true in the north Midlands where the sandy beds are thicker. It is necessary, therefore, to regard the term New Red Sandstone as a convenient label for a group of rocks of similar age, but to realize that all the rocks contained in it are not sandstones and that some are not even red; the geologist's habit of using such descriptive names in a chronological sense is sometimes misleading to a beginner. These rocks are called 'New' to distinguish them from the Old Red Sandstones, a group of somewhat similar character, which also is often red, but likewise is not wholly a sandy series. The newer group appear above the Carboniferous rocks (including the Coal Measures), while the older group in places forms their floor, and so the distinction was a matter of great importance in the early efforts to understand the distribution of coal.

The redness of these rocks is ascribed to their having been deposited under 'continental' conditions: in few cases have any of these beds been laid down in the sea, where the iron compounds are generally reduced and give to the deposits a blue or grey colour (as in the Lias clays or the Oxford Clay, which are typical marine muds): the state of oxidation of the iron in the New Red Sandstone, leading to the red colour, implies deposition under non-marine conditions. There are reasons to believe that some at least of the New Red Sandstone rocks were laid down when Britain was practically a desert.

The Trent Valley

For many miles the Trent meanders in its wide flood plain, its alluvial tract being nearly two miles wide. Much of this area is liable to flood, save that at many points raised banks or levees line its course. Near Trent Bridge in the south of Nottingham the alluvial area, still known as the 'Meadows', was not built over until late in the nineteenth century, and has only been rendered free from floods by the erection of considerable embankments.

Neither of the other two large towns of this region, Derby and Leicester, is situated on the Trent, but both are on its tributaries at some distance from the main valley and just on the edge of a wide area of low country which is subject to extensive floods. Both of these places are built on the red marls and, as in Nottingham, red brick forms the main part of their buildings.

The middle Trent flows through a populous area with many industries and a large coalfield, and the river is by far the largest source of surface water. Hence we find spaced at intervals of ten or fifteen miles along the river large coal-burning, electricity-generating stations. They are here because of local coal and demand for power, and more particularly because of their voracious need for enormous quantities of cooling water. It is for the same reason that all British nuclear power stations save one are in coastal or estuarine positions.

Naturally there are few villages within the damp alluvial tract near the river, which is given over to meadowland, willows fringing the channel in many parts. The river meanders in broad curves, for the gradient is low throughout this stretch and as the water is so little above sea-level it is unable to do much further work in deepening its valley. It flows slowly, swinging from side to side and widening its valley and building up its alluvial plain.

In the course of time each bend of the river tends to become more and more a complete loop, until ultimately, during some time of flood, the meander is cut off, the river shortening its course and leaving the old bed as an ox-bow lake, much in the way that modern straightened roads leave portions of the former bends derelict: so the old bed becomes in time silted up and curved, shallow depressions on the flood plain mark former positions of the river.

The cutting across of meanders has of course resulted in transferring some area within the meander from one side of the river to the other, and where county and parish boundaries have been drawn along the former positions of the river a portion of a parish has frequently been transferred to the further side of the river.

Within the flood plain the surface of the alluvium is almost flat, but even apart from the meander depressions there are minor irregularities which cause some parts of the meadows to be less frequently under water than others. There are many places where the banks of the river have been raised above the general level of the plain owing to the greater amount of sediment dropped there during times of flood. These raised banks, some of which have subsequently been strengthened artificially, are known as river levees and so tributaries find some difficulty, once they are on the flood plain, in reaching the main river, which in building up its banks tends to pond back its tributaries. One stream near South Muskham flows for about five miles more or less parallel to the Trent before effecting a confluence. Other small streams draining the flood plain actually rise within the raised belt, and flow away from the river for some distance.

From time to time the river in its meanders impinges against the higher ground formed by the solid rocks fringing the flood plain; the rocks are generally red marls, and locally the river is cutting into them to form steep cliffs; the red tree-covered cliffs below Radcliffe-on-Trent afford one of the best examples of this erosion of the solid rocks, showing how the river is extending its flood plain laterally, and illustrating the way in which valleys are gradually widened. The steep, wooded bank below Clifton Grove, formerly one of the beauty spots of the region, shows similar features, with the Trent again hugging the right side of its valley.

Where the Trent comes so close to the higher ground afforded by the solid rocks, it has been possible for villages to grow up near the river. But most of the villages of the river area are situated not on solid rocks but on patches of gravel which are slightly raised above the river plain, and which gain proximity to the fertile meadows while they themselves have dry situations. Such villages occur at short intervals along the valley, at Shelford, Gunthorpe, Hoveringham and Bleasby, while Beeston had a similar position, but has outgrown both the site and its village character. Some of these gravel terraces represent earlier flood plains of the river, which had an expanse even wider than that of the present alluvium. In this feature also the Trent is typical of most big English rivers; it reached base-level, at which it built up wide plains, but changed conditions have enabled it to cut down through those plains and by its meanders to reduce them to more or less isolated remnants situated at intervals along the valley tract, sufficiently far above the river to escape its floods. More details of the formation of such terraces are given with reference to the Thames. Further downstream the villages of the lower Trent valley become more scattered owing to the paucity of gravel terraces for dry-point settlement. Below Gainsborough, where the river flows through a countryside more analogous with the Fens, the layout of the riverside villages is of particular interest. All of the main river and almost all its tributaries have been embanked by large, artificial levees which form the highest points in the valley. Thus, to avoid the flooding which occasionally inundates the low-lying fields, the roads follow the levees for considerable distances to link up the small, elongated villages, usually only one street wide, which perch precariously on these embankments. Burringham and East Butterwick, near Scunthorpe, are particularly good examples of these curious village plans.

This area is, like the Fens, the site of a former pro-glacial lake. The Humber and the Wash were blocked at one stage in the Ice Age by an ice sheet in the North Sea (Fig. 17). The Trent drainage with that of other rivers was therefore ponded up in what is known as Lake Humber, which stretched from near Nottingham and Grantham to near Leeds and York. This lake was connected with glacial Lake Fenland by the Witham gap, on which Lincoln stands (p. 40). We have also seen earlier how the Trent formerly drained through this gap to the Wash, so for these reasons it is hardly surprising that this is a low-lying, frequently flooded region today. The remnants of former marshes are currently found in Hatfield Moors and Thorne Moors, while the Isle of Axholme was undoubtedly correctly named centuries ago. Artificial diversions have been made of the Don and the Idle.

Charnwood Forest

Rising unexpectedly from the clay plain about six miles north-west of Leicester are the surprising hills of Charnwood Forest. Craggy summits of bare rock and bracken-covered slopes above wide expanses of woodland introduce into the gentle fertility of the Midland plain a type of scenery of so unusual a nature that the hills appear to be more elevated than their actual 800 or 900 feet. The barren soils of these hills and the surprising character of their summits at once proclaim that different types of rocks are introduced among the red rocks of the plain and by their hardness stand out to form the high ground (Fig. 21). Various types of rock are here. Granite-like rocks of igneous origin, formed by the consolidation of molten magmas at great depths, are quarried at Markfield and farther north at Mountsorrel, chiefly for road stones. In Swithland and other parts, slates of similar general nature to those of Wales were formerly extensively quarried, while other hills show rocks representing hardened deposits of ash and debris from ancient volcanoes: it must not be supposed that these are volcanic peaks, for they have no relation to the unknown craters from which the material was derived, and the hills owe their form primarily to the hardness of the rocks.

These rocks are obviously more closely comparable with those of many parts of Wales than with anything ordinarily met with in central or eastern England. They represent, in fact, some of the oldest rocks known anywhere in England. Similar rocks may underlie much of the Midland area at varying depths, but here they have been exposed by the wearing away of the red marls from round about them. They form a number of 'islands' or inliers in

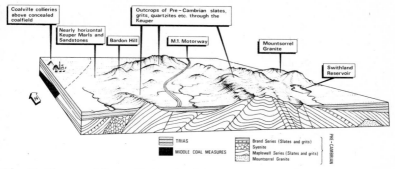

Fig. 21. *Charnwood Forest. The Pre-Cambrian rocks protrude from beneath the cover of newer rocks.*

the great stretch of marl (Fig. 21). They can best be visualized as a series of related and neighbouring mountain-tops which were buried by the red marl as it accumulated; the red marl was formed under desert conditions, part of it probably in drying up salt lakes comparable with those of south-western Asia, and the Charnwood hills stood high above these for ages, until the accumulating sediments slowly engulfed them. These red sediments were themselves buried under the Lias and later rocks when the whole region sank beneath the sea, but with its uplifting, long afterwards, the overlying beds have slowly been stripped off by rivers and atmosphere. At the present time just enough of the red marl has been taken from the area to expose the old mountain peaks. But as the red marl is so much softer than the old slates and granites, very little change has been effected in the form of these harder rocks during the removal of the marl.

As Professor W. W. Watts has suggested, therefore, Charnwood Forest presents us with an old landscape of hills and valleys which existed when the New Red Sandstone was formed, a landscape long buried and just being re-exposed as the red beds are removed. In the old surface there were several valleys between ridges running in a north-west to south-easterly direction. These old trenches have to some extent been re-excavated, and are followed by the older roads through the area, such as that from Shepshed up the Shortcliffe valley, and down Lingdale to Swithland, and that along the Blackbrook valley. Today, however, the more advanced technology of the road builders has enabled them to construct a motorway (M1) through the heart of Charnwood with little respect for its topography or geology. The main streams of the area flow along parts of the ancient valleys, but in parts of their courses they pass across the old ridges of hard rocks; the valleys then show striking contrasts, the streams having cut deep, picturesque gorges in the older rocks; the best examples are the Ingleberry gorge on the Shortcliffe brook, in the north of the forest, the Brand gorge about a mile south of Woodhouse Eaves, and the Bradgate Park gorge near Newtown Linford, at the southern end of the area.

To understand the development of these gorges and their relation to the wide valleys where the streams flow on the softer red marls, it must be remembered that the drainage of the area was initiated when the harder rocks were still completely buried. Streams running on the softer rocks in cutting down their valleys here and there came across projecting ridges of the old rocks, into which they continued to cut downwards. But they were unable to widen their valleys in these harder beds as they could in the marls, nor could they cut downwards so quickly, so that streams flowing on the marls filling the old valleys gained over those which had to contend with slates and granites; hence the main streams came increasingly to follow the original channels in the old rock surface.

The West Midlands

The remainder of the Midland plain is largely made up by the outcrops of the various divisions of the New Red Sandstone together with extensive areas of Coal Measures. These latter are mainly areas of clay and shale, giving rise in many parts to a dull grey soil associated with the familiar evidences of coal-mining, though red rocks are very widespread in the Warwickshire coalfield. The coalfield areas of both Warwickshire and south Staffordshire are rather more elevated than the surrounding plain, and the chief valleys therefore lie outside these regions, frequently on the red Keuper Marls, the Tame joining with the Anker at Tamworth where the castle guards the passage of the united rivers.

The borders of the coalfield plateaux are sometimes marked by a definite edge, especially at Nuneaton, on the north-east of the Warwickshire coalfield. Here ancient rocks, including some of volcanic origin not very different from those of Charnwood, make up a sharp ridge facing across the marl and clay country which stretches through the west of Leicestershire as far as Charnwood Forest. Similar ancient rocks give rise to the Lickey Hills some ten miles south-west of Birmingham.

But apart from these occasional inliers within the Midland plain, the greater part of the area is made up by the New Red Sandstone, which gives rise to low, marly country or to rather more elevated and drier tracts according to the distribution of the marls and sandstones. Narrow outcrops of Bunter and Keuper Sandstones stretch north and south of Birmingham, forming a ridge on which the city is built overlooking the coalfield of the 'Black Country' of south Staffordshire. Dry, sandy soils formed on the Bunter Sandstone were among the last to be cultivated, and some heath and natural woodland may still be found in Sutton Park on the fringe of the built-up area. On the Keuper Sandstones the soils have neither the heaviness of the marl regions nor the dryness of the Bunter, and give rise to a fertile belt which attracted early settlements at Bromsgrove and Sutton Coldfield. Here the purple and red sandstones have been used in building, but over much of the area red brick predominates, and most of the conurbation is characterized by those rows of dull brick houses which are the mark of industrial progress in most parts of the Midlands.

North Staffordshire

Lying at the north-west corner of the Midland plain the north Staffordshire coalfield stands on an important watershed between the rivers of the Midlands, notably the Trent, and the rivers which flow across the Cheshire plain to the Mersey. Although the New Red Sandstone rocks pass northwards into Cheshire and south Lancashire this northern extension of the Midland plain is narrowed by the low hills on which the Pottery towns are built. Although these hills are geologically part of the Pennine foothills, the character of the Potteries and the surrounding countryside is linked essentially with the Midland plain.

Travelling north from the 'Black Country' of south Staffordshire we pass the city of Lichfield, with its triple-spired cathedral of red sandstone, and the rapidly expanding county town of Stafford on the flanks of Cannock Chase. This group of hills is carved from Bunter Pebble Beds and Sandstones which produce only poor stony soils given over mainly to heathland and forestry plantations. Farther to the north the Bunter Pebble Beds and Sandstones appear again and the scarp formed by these creates a conspicuous ridge, running in an arc northwards from Stone to beyond Whitmore. The vegetation of this ridge differs little from that of Cannock Chase, for large areas of heathland and coniferous forest flourish on the red sandy soils. The Keuper Sandstones are near enough, however, to provide building stones for the churches and the more eminent secular buildings of the district, especially at Trentham and in the villages such as Maer and Ashley which lie to the south of the wooded ridge, where the Keuper Marls give to the landscape the typical pattern of Midland farmland. If we cross the Shropshire border here the half-timbered towns and villages, which characterize this part of Midland England, remind us that the Welsh border country is not far to the west and that the 'gap' between the North Staffordshire hills and the Shropshire hills is less than twenty miles in width. The routeways from the Midlands to north-west England have reflected the importance of this gap (sometimes called the 'Cheshire Gate') from earliest times, but it was after the Industrial Revolution that this part of England became of greater importance. Thus we see the construction of the Grand Trunk Canal, soon to be followed by the Shropshire Union Canal, linking four of the largest English rivers (the Trent, Mersey, Dee and Severn). Not long afterwards came the railways, with the line from London to Crewe taking advantage of a glacial meltwater channel which breaches the Bunter scarp at Whitmore. Finally the M6 motorway has come this way too, not as dependent on the geology and topography as the canals and railways, but still using glacial meltwater

Fig. 22. *The north Staffordshire coalfield (after S. H. Beaver).*

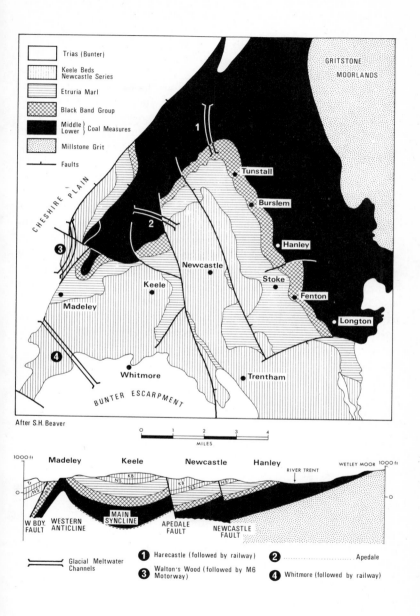

Trias (Bunter)

Keele Beds
Newcastle Series

Etruria Marl

Black Band Group

Middle } Coal Measures
Lower }

Millstone Grit

Faults

GRITSTONE
MOORLANDS

CHESHIRE PLAIN

1

Tunstall

Burslem

Hanley

2

3

Newcastle

Stoke

Keele

Fenton

Madeley

Longton

4

Whitmore

Trentham

BUNTER ESCARPMENT

After S.H. Beaver

0   1   2   3   4
MILES

1000 ft   Madeley   Keele   Newcastle   Hanley   WETLEY MOOR   1000 ft
RIVER TRENT

0   0
Tr
NS
KB
KB
NS
NS
W BDY.   WESTERN   MAIN   APEDALE   NEWCASTLE
FAULT   ANTICLINE   SYNCLINE   FAULT   FAULT

Glacial Meltwater
Channels

1 Harecastle (followed by railway)   2 .... Apedale

3 Walton's Wood (followed by M6
Motorway)   4 Whitmore (followed by railway)

channels to cut through the hard rocks which form the western rim of the north Staffordshire coalfield at Walton's Wood. north of Madeley (Fig. 22).

The Bunter escarpment of north Staffordshire is an important boundary in the landscape of this area, for it divides the good farming land of the Keuper Marl country to the south from the thinner, poorer soils of the coalfield to the north. As we pass from the Bunter hills northwards at first the colour of the soils and the elevation of the countryside change but little, for the hard, red Keele Sandstones here represent the upper part of the Coal Measures and have been utilized as an important building stone in the local churches, including the parish church of Newcastle-under-Lyme. But it is the Coal Measure marls (known as the Etruria Marls) which lie between the Keele Sandstones and the underlying coal-bearing rocks which have played so important a part in the history of north Staffordshire and in the creation of the unique Potteries landscape. Although these clays had long been used for local pottery manufacture it was not until the mid-eighteenth century that their wholesale exploitation saw the rapid growth of the six Pottery towns of Stoke-on-Trent with their curiously shaped bottle-ovens. The siting of the six towns is of special significance for they each straddle the junction of the Etruria Marls and the coal-bearing series of the Middle Coal Measures, for, as Professor Beaver has pointed out, the supply of coal was equally important since it took six times as much coal as clay to produce a given quantity of earthenware (Plate 16). Today the local clays are no longer used for earthenware manufacture but only for coarse drain-pipes and tiles, but the legacy of the industry has left a forlorn landscape of gaping marl holes, mountains of broken crockery (shraff heaps) and derelict factories. All this in association with the usual colliery clutter and smoke-blackened brick terraces of a typical coalfield settlement creates a scene of spoliation rarely equalled in the English countryside. It makes it all the more surprising therefore to find that only a few miles outside the city limits, as we pass across the boundary of this small coalfield, the rural scenery is virtually untouched: the wild. beautiful gritstone moors to the east and the rich farmlands of the Cheshire plain to the north and west.

Plate 16. *A landscape in the Potteries, north Staffordshire. The pottery towns are located at the junction of the Etruria Marls and the Middle Coal Measures. In the foreground a marlhole is being actively worked to supply the roofing tile manufactory seen in the middle distance. Beyond is the spoil heap and winding gear of a colliery which supplies the coal for the ovens. The well-known bottle-ovens have now almost gone.*

The Cheshire Plain

The red marl country covers much of Cheshire and north Shropshire, though this plain is also fringed by outcrops of the sandstone beds. In its main features the scenery is thus closely comparable with that of the Midland plain. The dull red brick town of Crewe, which has grown up as a railway centre where not even a village existed before, may be contrasted with old Chester, with its half-

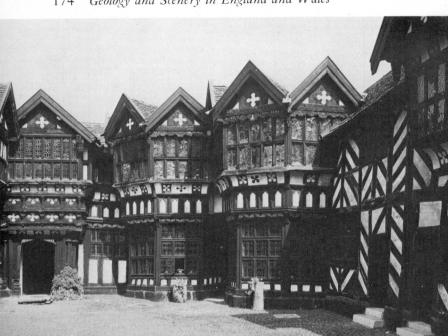

Plate 17. *Little Moreton Hall, Cheshire. The 'magpie' architecture is characteristic of the Heart of England, especially on the Keuper clay plains or the drift-covered lowlands of Cheshire and south Lancashire. In regions removed from supplies of suitable building stone the domestic architecture frequently exhibited extensive half-timbering before brick was more widely introduced. Note the use of Bunter pebbles to surface the courtyard.*

timbered houses and stone city walls; Chester Cathedral is a notable example of the use of the warm-toned Keuper stones, which are so extensively used in church construction throughout the county. At a few places such as Marton and Warburton, however, even churches are half-timbered.

Although the Cheshire lowlands are invariably referred to as the Cheshire plain there are in reality two separate plains, representing the drainage basins of the Dee and the Weaver/Dane respectively. These are divided by a conspicuous ridge of Keuper Sandstones which runs north from near Malpas past Tarporley to the imposing red cliffs of Helsby which overlook the Mersey. The northern end of this ridge was formerly covered by the extensive Delamere Forest, but this has dwindled in size throughout the centuries. Here was one of the great sources of raw materials for the domestic architecture of the plain. the attractive black and white half-timbering of Elizabethan England perhaps reaching its finest development in towns like Nantwich or in individual buildings such as Little Moreton Hall, to the south of Congleton (Plate 17).

At the southern end of the mid-Cheshire ridge, with its castles and churches of pink Keuper Sandstone, the Cheshire–Shropshire border follows a less conspicuous ridge which runs eastwards from Ellesmere, through Whitchurch, before impinging on the hills of the North Staffordshire coalfield at Madeley. At no point on this broad but subdued ridge are solid rocks visible at the surface, although it is a major drainage divide between the Severn and the Dee/Mersey. The latter ridge is in fact a very recent addition to the geological scene, for it represents a vast recessional moraine deposited by an ice sheet which formerly occupied the Cheshire basin. The extensive gravel and sand deposits of Cheshire, the quarrying of which creates such conspicuous intrusions in the green chequerboard field pattern, are the legacy of this vast ice sheet.

One other feature of the scenery of Cheshire may be noticed in the existence of wide lakes, formed as a result of salt working. In the red marls salt occurs in beds with an average thickness of about 100 feet, formed by the evaporation of the salt lake in which much of the Keuper Marl was deposited. The extraction of these thick beds has caused the surface to subside to a considerable extent. Subsidence is of course frequent in coal-mining areas, but it must be remembered that coal seams are usually no more than a few feet thick, and that supporting pillars and packing prevent collapse to a certain extent, while in working salt by pumping it out in solution the whole of the bed is removed and no support is left. So in many parts of the salt-mining area, between Nantwich and Northwich, great meres or flashes are common, and are constantly extending.

11. Mountain Limestone Country

It has been seen how the scarplands of south-east England are composed, for the most part, of a variety of limestones, from the Lias, through the oolites, Corallian and Portland Limestones of the Jurassic to the Cretaceous Chalk. In certain regions these rocks form hills with steep slopes, especially on the scarp faces, but even if we stand on such abrupt eminences as White Horse Hill on the Berkshire Downs, or Cleeve Hill on the Cotswolds, there is scarcely any bare rock to be seen. Except on the coasts where many of these Mesozoic limestones form precipitous cliffs the downlands and wolds are devoid of crags because these rocks all break down fairly readily, their hills become grassed over, and their steeper slopes are often given over to woodland.

To find inland crags and cliffs of bare rock we have to travel into 'Highland Britain', farther north and west of the Midland plain, and we have in general to seek rocks older than the New Red Sandstones.

In England no type of rock gives rise to this exciting country more frequently than the Mountain Limestone or the Carboniferous Limestone (the latter name because it is found in close association with the Coal Measures, which it underlies). Unlike the Chalk and Jurassic limestones the Mountain Limestone is not found in continuous belts running right across England; its outcrops are practically limited to the west of a line from Lyme Regis to Whitby, but they are more complex than those already dealt with. Briefly, the reason for this is similar to that given for the comparatively complicated pattern of the chalk outcrop, namely, folding of the beds followed by the wearing away of the tops of the anticlinal folds. But the Mountain Limestone in places has been folded much more severely than the Chalk or any of the newer rocks, for it was involved in Hercynian mountain-building movements before the New Red Sandstone or any later rocks were formed, and in many places it shows none of that simplicity in broad structure which characterizes the newer strata; frequently it is buried by these newer rocks.

It is not surprising, therefore, that the Mountain Limestone forms discontinuous outcrops in many areas; in the Mendip Hills and many other hill masses around Bristol; along the Gower coast of Glamorgan and round the edges of the South Wales coalfield; at Chepstow and along the borders of the Forest of Dean; in the Peak District, and around Settle and Skipton in Yorkshire and farther to the north; in the Great Orme at Llandudno and near Llangollen on the Welsh border. This limestone is a widespread formation, and over much of England and Wales it retains similar characters, which will be recognized by those who are acquainted with any of these areas just mentioned. In all of them there are rocky gorges with steep or even precipitous sides of bare grey limestone, magnificently seen in the Cheddar Gorge and in the gorge of the Avon at Bristol, in the Wye valley above Chepstow, and so on through all the regions named.

As seen in any of these areas the Mountain Limestones occur in thick beds, individual beds often being ten feet or more thick so that the exposed rocks appear massive and compact, differing in these respects from many of the limestones of the Cotswold stone belt, and for this reason forming bolder country. The beds, moreover, are often tilted and bent into great folds (especially in the south-west) and have clearly been disturbed greatly since they were formed as horizontal layers beneath the waters of a clear sea, at which time all the areas referred to were at one level and were continuous. The changes that have taken place since the time of formation of the limestones will be more easily comprehended when it is remembered that they are older than any of the rocks so far mentioned in this book (except those of the Lickey Hills and the Charnwood Forest); not only has the sea in which they were deposited no relation to any existing seas, but much of England and Wales has been under an extensive sea on at least two subsequent occasions. In short, when the Mountain Limestone was formed scarcely any part of England existed. Since that time the rocks laid down on the floor of the Carboniferous sea have been heaved up to form the tops of present-day hills (and of much higher mountains that have long been worn away) and depressed in other places far below sea-level.

These limestones were obviously formed in a sea, for in many beds fossil shells and corals are common, as well as the remains of other less familiar creatures. The limestones in fact are almost entirely composed of carbonate of lime, derived from the remains of skeletons of marine invertebrates, very little material from any inorganic source being present in some beds: it follows that little sand or mud was then being carried into this part of the sea, and the water must have remained clear. Occurring among the limestones are some layers of shale, which frequently become important near the top of the group and in places also near the base, but for the most part the 3,000 feet of strata consist of limestones which from the scenic point of view are of similar character, almost universally grey in colour, tending to appear black in some areas or white in others, though they are also occasionally stained red.

These Mountain Limestones are cut by well-developed joints nearly always at right angles to the bedding. Along these joints the rocks tend to break cleanly into blocks giving clean faces, and in many places the joint faces determine the precipitous character of mountain-sides and gorges which is so striking a feature in all areas of Mountain Limestones. The great cliff of High Tor, seen in the familiar view along the Derwent at Matlock, is a splendid example, but the cliffs of the Avon Gorge and of the Wye are equally impressive.

Like all limestones, the beds of this group are soluble in water containing a little carbon dioxide, so that rain water (which normally takes up a little of this gas on its passage through the atmosphere) is capable of dissolving the limestone on which it falls. Owing to its purity, that is its freedom from non-calcareous and insoluble material, the solution of the surface of the Mountain Limestone by atmospheric water gives rise to very little soil, a condition already noticed in the case of the Chalk. Thus in some

Plate 18. *The bare limestone pavements above Malham Cove. The clints are clearly shown in the foreground while the horizontal bedding of the Mountain Limestone can be distinguished in the cliffs of the cove itself.*

areas the limestone gives rise to great barren tracts devoid of any plants except such as have been able to secure a foothold in the crevices in the bare rocks. Such conditions are naturally best found on high ground, where there is no opportunity for soil to accumulate from higher levels, and where the rain water gradually widens the joints into which is washed any soil that may be formed.

No areas better illustrate these conditions than the great 'karst' plateaux around Ingleborough and above Malham Cove near Settle (Plate 18). Here are wide, level stretches of bare, light-coloured limestone, with the joint planes enlarged by the solvent action of rain so that wide, irregular chasms (known as clints or grikes) trench the surface, the channels varying in width up to several feet and being of considerable depth. Here and there a bush of hawthorn contrives to exist on this upland, while in the shady crevices are the hart's-tongue fern, wood sorrel, and other plants. In some of these areas there is heavy rainfall, but all the water rapidly passes underground. Likewise on the more extensive moorlands where grassy pastures are typical, as in the Craven Hills north of Skipton and in many parts of the Peak, the ground is generally dry, but a varied flora thrives, often with wild thyme and the yellow mountain pansy in abundance. On these uplands are stone walls and houses reminiscent of the Cotswolds, but the harder limestone has not favoured such ornament or elegance in building as is characteristic of freestone regions, and houses are solid, grey and cold-looking.

On the limestone uplands, especially in Yorkshire and Derbyshire, farms are chiefly occupied with sheep grazing; probably the grazing restricts the growth of trees. But where the Mountain Limestone is carved into valleys, trees are more frequent, especially on the valley slopes. On these steep hillsides the soil is typically shallow, so that oak is excluded just as it is from the soils of the steep slopes of the chalklands. The terrain of the Mountain Limestone is more irregular than that of the Chalk, and because of its jointing and comparative hardness large, blocky screes characterize the foot of the limestone cliffs. It is upon these talus slopes that the ashwoods flourish although other lime-loving shrubs, such as hazel, hawthorn and elder, can also be found. But since ash is practically the only timber tree of the Derbyshire dales it has been severely exploited in the past. The Derwent valley along the east of the Peak District is well wooded for many miles, and in numerous regions the steeper parts of the limestone outcrop are marked by woodlands. Rarely is there any arable land on these areas, for the soil is ordinarily too thin to allow of ploughing: different conditions may be introduced, however, by a cover of superficial gravel or clay.

The solubility of the limestone and the tendency for the joints to become enlarged have led to another important feature in its scenery, for here, as in other limestone regions such as the oolites and the Chalk, much of the drainage tends to flow underground,

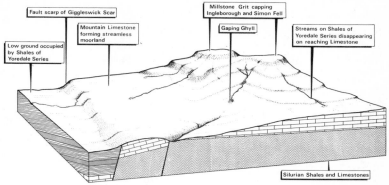

Fault scarp of Giggleswick Scar

Mountain Limestone forming streamless moorland

Low ground occupied by Shales of Yoredale Series

Millstone Grit capping Ingleborough and Simon Fell

Gaping Ghyll

Streams on Shales of Yoredale Series disappearing on reaching Limestone

Silurian Shales and Limestones

Fig. 23. *The country near Ingleborough, viewed from the south-east.*

its passage in the case of the Mountain Limestone being generally through crevices dissolved out and enlarged to form caves. Thus on many mountains of this type the higher valleys are normally dry, or perhaps are occupied by streams during very wet seasons only: most of the water passes underground, sometimes unobtrusively by percolation, sometimes by large and definite passages or swallow-holes. The swallow-holes of the Yorkshire hills are well known, the great chasm of Gaping Gill (or Ghyll) on the east of Ingleborough being one of the most impressive (Fig. 23). It is a vertical shaft over 350 feet deep, opening out into an enormous cavern into which falls the highest known waterfall in England, the water arriving from another underground channel. The caves in the limestone worn by these underground streams scarcely enter, perhaps, into the category of scenery, but they are familiar attractions in all such limestone areas. Possibly they are best known to the ordinary traveller in the Mendip area where at Cheddar and Wookey Hole the caves are daily thronged throughout the summer with visitors from all over Britain. In these caves another aspect of the work of underground water is made evident, in the deposits of carbonate of lime both in the form of stalactites and stalagmites, and in the more irregular spreads of tufa, material which has been dissolved at one place being set free again in others when conditions decrease the solvent power of the water.

The Mountain Limestone of West Yorkshire

The limestone uplands of Yorkshire are more extensive if less accessible than those of the Mendips, and while perhaps there are no
gorges finer than that of Cheddar the great tracts of limestone of
the north may most suitably be chosen as illustrations of the
scenery formed by these particular rocks. The limestones occupy
large, but somewhat irregular, areas in the Pennine uplands below
the summits of Whernside, Ingleborough and Pen-y-Ghent. These
areas are best reached from Skipton and Settle. On the plateau
under Ingleborough, the features of the limestone scenery are
extremely clear and call for little explanation: the top of this,
possibly the most interesting mountain in Yorkshire, is formed by
nearly horizontal grits overlying shales with thin beds of limestone.
An outlier of newer rocks, it stands out above the griked limestone
surface, and down the steep mountain-sides flow numerous swift
streams (Fig. 23). As these reach the more level limestone surface
they plunge underground, Gaping Gill on the south of the mountain being one such swallet, Alum Pot on the north another. These
streams flow underground for some distance, reaching the surface
again either when they leave the limestone outcrop or when the
valley is cut down below the level of saturation. The identification
of each particular stream at its outflow by the introduction of
chemical dyes into the water going underground has made possible
the mapping of the direction of the underground courses.

The rivers flowing west from Ingleborough rise to the surface
when they reach the base of the limestone, where in Chapel le Dale
it rests on the impermeable and insoluble rocks which lie beneath it.

All the country hereabouts shows features equally striking. Above
the village of Malham, some five miles to the east of Settle, is found
unrivalled limestone scenery (Plate 18). At Malham Cove is an
outflow of water, a small stream which ultimately becomes the
Aire; the spring occurs at the foot of a great limestone precipice,
but it needs little imagination to realize that the gully above the
cliff is the natural bed of a stream. Formerly it flowed across the
high ground above the precipice from its source in Malham Tarn,

making a great waterfall nearly 300 feet high, but gradually the water has found a new path underground beneath the plateau, and no water now dashes over the fall, although the change appears to have taken place fairly recently, for there are records of water coming that way during times of flood within the last century. Now no flood yields water enough to exceed the capacity of the underground channel.

Only a short distance away is Gordale, one of the finest lime-stone gorges in the country. It is probably a cavern of which the roof has collapsed. The walls are steep and mostly bare, the clean joint planes making great precipices where they cut through the massive beds; in several places the walls overhang slightly and at one point a remnant of the roof still remains. Near this is a small waterfall, for some water still flows along the surface, though the supply is much reduced owing to the fact that the greater part of the stream sinks below ground some way up the valley, to reach the surface again near the mouth of the gorge. Here then is an old cave which has lost its roof and has almost lost its water-flow; the stream first passed underground to form the cave which is now represented by the gorge, and later found a still lower path. So the water is continually being 'short-circuited'. Waters in limestone regions continue to make lower courses, leaving one cavern dry when they have formed another, until they are at last held up by impermeable beds, temporarily if the beds are only thin and are underlain by more limestone, permanently if they have reached the real base of the limestones, for example the Greta and its tributaries.

In the area near the Yorkshire and Lancashire borders, another type of feature is shown. Running in a straight line from north-west to south-east immediately above the road from Settle to Ingleton is a great wall of limestone. This scarp, known as Giggleswick Scar, is formed by bare, grey limestone, and rises above a wide tract of more varied character made up of grits and shales (Fig. 23). Here is a feature quite unlike anything which has been discussed previously, for it differs from the scarps of southern England not only in its straightness but in the fact that the rocks occupying the low ground do not pass underneath the limestone; in fact the limestone is present at some considerable depth, underneath the rocks forming the lower ground. The beds forming the slopes of Ingleborough, many hundreds of feet above the limestone scarp, are represented at some depth in this low ground (Fig. 23). This scarp marks the course of a great fault or dislocation in the rocks, a complete break in the areas already described, but they are not frequent, and are without reference to breaks in strata. Such breaks are not unknown in the areas, already described, but they are not frequent, and are generally inconspicuous in their scenic effects.

In Giggleswick Scar, however, the fault scarp is as impressive as any in England, a great break in which the ground on the south-west has fallen relatively to that on the north-east. It has generally been supposed that although the present cliff is along the line of the fracture, it does not represent the step produced when the break occurred, and that possibly great thicknesses of rock then covered both areas. It has been thought that the limestone edge stands out as a scarp because the fault brought the hard Mountain Limestone on a level with much softer shales and sandstones.

In the area around Settle are several similar faults, running more or less parallel to one another and doubtless caused by similar earth stresses. These form the Craven faults, but the others do not give rise to any scarps so prominent as that of Giggleswick Scar, and in many places the scenic effect is very slight, the upland surface passing across the fracture line with little change of level, so it is believed that these faults existed long before the Tertiary movements but were not affected by them.

The Knoll Country

To the south of this region and north of the margin of the coalfield of Burnley and Accrington is another great expanse of Mountain Limestone, occupying the Craven country between Settle and Burnsall and Clitheroe on the Lancashire border. These regions are not so high generally as the areas just described, and the limestones are interbedded with much shale. They are also peculiar in that there occur a large number of queer mound-like hills rising abruptly for as much as 200 feet above the surrounding country; these hills are often conspicuously dome-shaped, but vary somewhat in form

Plate 19. *A Mountain Limestone Reef Knoll. The highest point is Chrome Hill, near East Sterndale, Derbyshire, and is formed from the reef limestone. The more sloping ground in the left middle distance is made up of shales of Millstone Grit age.*

(Plate 19). Occasionally they are over half a mile across at the base, but small knolls only a hundred yards across also occur. They are seen in Burnsall and Thorpe in Wharfedale, and along the road by Cracoe towards Skipton; near Clitheroe a line of knolls runs east towards Twiston, including Coplow and Salt Hill, just north of Pendle Hill. Such knolls are known in other parts of the north of England in Mountain Limestone tracts, but it is a type of scenery most familiar in this area. There has been much discussion regarding the origin of these extraordinary mounds, some believing that they have been produced by movements in the rocks after they were formed, others that the knolls are due to some original peculiarity of the deposit: in many knolls the limestone is exceedingly rich in fossil shells. It is now considered that these are 'reef' knolls with some special condition of deposition in that part of the Carboniferous sea being responsible. Each dome corresponds to a 'reef' deposit of more resistant character and the removal of the surrounding softer rocks has left it projecting.

The Peak District

Another great area of Mountain Limestone forms the High Peak, the most southerly part of the Pennines, to which it is desirable to make some further reference since, quite apart from its special attractiveness, its characteristic features are more easily accessible than many of those of the Skipton and Clitheroe areas, for on its margin are such resorts as Matlock and Buxton. These spa towns, like Bath, owe their development to the exploitation of the natural hot springs which were utilized as early as Roman times. The mineral waters are obviously related to the lead and zinc lodes which occur extensively in the limestone and which are known to have been worked by the Romans. The mineralization is probably due to an intrusion of igneous material at depth, not yet uncovered by erosion, although related to this buried mass are the dark-coloured Derbyshire 'toadstones' which are in reality basaltic lava flows interbedded with the limestones. In the country west of Matlock and again between Tideswell and Buxton, the uniformity of the Mountain Limestone is broken by the presence of these dark rocks. Basalt is a volcanic rock, different in all its characters from sedimentary rocks which make up most of the country already described: it consists of a fine mass of crystals of minerals (olivine, feldspar and augite), formed by the rapid cooling of molten material. The basalt represents a lava poured out on the floor of the sea, which spread over a wide area before it became solid, when deposits of limestone were laid down above it. Evidence of volcanic activity is not of frequent occurrence among rocks of this age in England and these areas afford the best examples.

As these beds are not of great thickness their scenic effects are slight, the dark lavas being traceable here and there on the face of lighter limestone precipices, while by their impermeability they hold up the underground waters in a few places. But in several localities the volcanic rocks give rise to more remarkable features, for small, rounded hills at Castleton, Grange Mill, Hopton and Kniveton Wood mark the position of the vents or pipes of ancient volcanoes from which the lava was ejected. These hills are composed of volcanic debris, the hardness of which leads to them projecting somewhat above the level of the surrounding limestones. Those at Grange Mill, about five miles to the west of Matlock, are quite conspicuous and are typical of some others: two grassy, dome-shaped hills, the smaller only 300 yards across and rising a little over 100 feet, differ in contour and in colour from the craggy limestones around. Another vent occurs about a quarter of a mile west of the entrance to the Peak Cavern at Castleton.

Among the characteristic features of the limestone country in the Peak District are the underground streams and great caverns. The larger and deeper dales, however, descend below the water-table of the limestone and support perennial rivers which flow in places beneath sheer cliffs or fantastic crags of white and grey rock rising abruptly above the trees. All these features are seen in Dove Dale where Izaak Walton described the beauties of the gorge '. . . while seeking to beguile the wary trout . . . or scaling the steep ascent to Reynard's Cave, or standing with wondering admiration before such mighty monoliths as that called Ilam Rock, or the Pickering Tor'.

In shape this limestone tract is half an ellipse, and this form reflects its broader structure, for the region is essentially a denuded anticline, the centre of which is known as the Derbyshire Dome, partly buried under the New Red Sandstone of the Midland plain near Ashbourne, a small market town on the borders of these two regions (Fig. 24). On the western margin at Buxton the limestones dip to the west under the Millstone Grit moors of Axe Edge, while on the east at Bakewell the dip is easterly under the similar moors of south Derbyshire. From some of the higher hills it is possible to see the steep scarps of these grit moors bounding the limestone tract both on the east and west. Between the limestone and the grit edges occurs the outcrop of the shales already referred to as

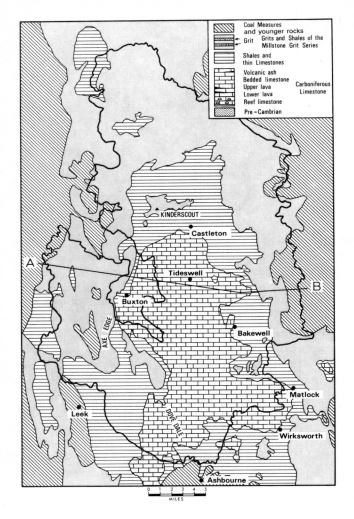

Coal Measures and younger rocks

Grit — Grits and Shales of the Millstone Grit Series

Shales and thin Limestones

Volcanic ash
Bedded limestone
Upper lava
Lower lava
Reef limestone — Carboniferous Limestone

Pre–Cambrian

KINDERSCOUT

Castleton

Tideswell

A

B

Buxton

AXE EDGE

Bakewell

Matlock

Leek

DOVE DALE

Wirksworth

Ashbourne

0 1 2 3 4 5
MILES

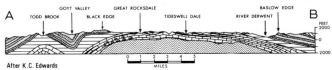

A GOYT VALLEY GREAT ROCKSDALE BASLOW EDGE B

TODD BROOK BLACK EDGE TIDESWELL DALE RIVER DERWENT

FEET
2000
0
2000

After K.C. Edwards

0 1 2 3 4 5
MILES

occurring above the limestones (which it will be convenient to refer to as the Yoredale beds), and as might be expected this belt has fixed the positions of the more important rivers, the deepening of the shale outcrop leading to the etching out of the harder bands.

Along the eastern outcrop the shales are followed for a great distance by the Derwent, its wooded valley overhung by the continuous wall of the unbroken grit scarp. From that eastern side it scarcely receives a tributary of importance, though it is joined by several rivers from the limestone area on its right bank. These rivers, the Wye the largest among them, flow through limestone dales; and 'the whole glory of the country is in its dales' where visitors marvel at the stately homes of Chatsworth and Haddon Hall. These, like The Crescent at Buxton, were built from the Kinderscout group of the Millstone Grit, one of the finest building stones in northern England.

Most of the upland valleys are dry during many months of the year, for much of the water is underground. Professor W. G. Fearnsides believed that 'the slitting of the lead veins in the limestone country has opened easy paths for the descent of surface water, and the adits to the mines have permanently lowered the water-table in their vicinity'. During the later part of the eighteenth century drainage tunnels or 'soughs' were driven from the Derwent valley westwards into the country where lead-mining was in progress: not less than thirty miles of such channels are operating, and result in bringing the water-table in the area down to the level of the main river. Consequently many of the higher dales are dry. Although they do not greatly disfigure the landscape the abandoned lead mines are a feature of the limestone country of the Peak District. At the height of its prosperity the area produced some 10,000 tons of lead ore annually, but working has now ceased although the estimated reserves are great. Some of the larger 'rakes', as the workings were called, have been shielded from view by long, narrow belts of woodland, partly to screen the disfigurement but also to prevent grazing animals from straying into the workings and being endangered by lead poisoning. The greatest

Fig. 24. *The structure of the Peak District. The thick black line marks the limit of the National Park.*

impact on the natural landscape is from the extensive limestone quarries at Miller's Dale, Matlock, Wirksworth and Hope Dale, which, because of their unsightliness, have been excluded from the National Park (Fig. 24).

The Derwent valley from Rowsley almost to Matlock is relatively wide, but just above Matlock it enters a narrow gorge cut in Mountain Limestone, and flows between the picturesque precipice of High Tor and the wooded slopes of Masson. From some viewpoints near the summit of High Tor the significance of this gorge is apparent; the course of the Derwent, following the narrow outcrop of the Yoredale beds, brings it against a projecting tract of limestone, thrown out of its usual position by minor folding. A slight detour to the east would enable the river to continue along the shale belt, which gives rise to a noticeable depression behind High Tor, but the river keeps to the limestone until it reaches Cromford, where it returns to the shales. At Cromford the river Derwent was utilized in the late eighteenth century by Richard Arkwright to power his cotton mill, the first example of such use in the cotton spinning industry. The Arkwright mill still stands, to remind us that during the Industrial Revolution, because of the availability of water power, many mills sprang up on the banks of the Derbyshire streams, and for a time the cotton industry here equalled that of Lancashire itself.

On the other side of the great anticlinal fold which forms the Peak District the river Dove for some time occupies a corresponding position on the western shale outcrop, but like the Derwent it also turns on to the limestone tract above Ashbourne, cutting there the wonderful, wooded gorge of Dove Dale. Unlike the limestone area near Settle, these rivers have nowhere cut down to the bottom of the limestone, and no older rocks are seen in Derbyshire.

Between these deeply cut valleys the grassy limestone uplands reach a height of well over 1,000 feet. But it is scarcely the bleak moorland which might be expected at this altitude, and there is much pastureland. Seen from above, the valleys often seem mere scratches in this plateau surface. Since the valleys are so narrow the chief routes have crossed the uplands themselves, and the Roman road follows the crest from Wirksworth to Buxton. Many of the villages are found up on the high land, generally constructed, as are the field boundary walls, from crudely dressed blocks of limestone; the dwellings sometimes are covered with a lime wash or a rough cast of limestone chippings. The grey-white limestone walls contrast sharply with the brownish tone of those made from the shaly grits and the dark brown of the weathered gritstone blocks. Where both limestone and gritstone are available near by the more easily dressed grit is generally utilized. Where the gritstone quarries are too far distant, however, limestone is used, as in the group of villages that include Tissington, Thorpe and Parwich which are located on a broad outcrop of shales. Flagstones, which occur within the Millstone Grit, were long used for roofing in the Peak District, and all door-posts, lintels and gate-posts of the older buildings were made from gritstone owing to the dearth of timber on these bleak uplands.

A final word is necessary on the relationship of the village settlement pattern to the geology of the area. Because of difficulties of water supply the villages of the limestone country are compact, sometimes following a linear plan along the narrow floor of a dry valley, as at Taddington, Eyam and Chelmorton. Where lead-mining intervened this linear pattern was lost, as at Bradwell and Winster. The settlement pattern on the grit and shale country, however, is very different, for here water supply is less difficult and the settlements are scattered instead of clustered.

Away to the east of the Peak, limestone again reaches the surface in small areas at Ashover, and at Crich. Both these areas are surrounded by extensive tracts of grit and represent small, subsidiary, dome-shaped anticlines where the erosion of the higher beds has exposed patches of the limestone.

No special reference is made here to the Mountain Limestone areas of South Wales and the Bristol district, since these are more appropriately described in later chapters. They show underground streams, steep-sided gorges, bare rocky uplands, and other features comparable with those already described. In these areas, however, the rocks are often more sharply folded or more steeply dipping, so that they rarely occupy such wide areas as in the north, although they represent no less variety of scenery.

12. The Pennine Moorlands

The mountainous country which forms the axis of northern England includes many limestone regions of the type referred to in the previous chapter, but there are even greater tracts of dark, barren moorland, covered by heather and by peat bogs, and it is this type of landscape, more than any other, which typifies the Pennines. The moorlands generally occur wherever the rocks known as the Millstone Grit (locally, Moorstone Grit) appear, but are also found extensively on the somewhat older group of Carboniferous sandstones, limestones and shales which are grouped together as the Yoredale beds. Since we have already described the Mountain Limestone terrain of Craven and the Peak District in the previous chapter we shall concentrate mainly on the areas formed from these other Carboniferous rocks. Despite the ubiquitous moorland scenery of the Pennines, structurally the region may be split into two parts, the Southern and the Northern Pennines, with the boundary occurring near to Skipton. Each of these divisions will be further subdivided to facilitate description: the Southern Pennines consist of the Peak District (which includes the Derbyshire Dome) and the Central Pennines; the Northern Pennines can be split into the Askrigg Block, divided from its northern neighbour, the Alston Block, by the Stainmore Gap (Fig. 25).

In the Peak District of the Southern Pennines the highlands have been carved from a fairly symmetrical upfold or anticline which has resulted in the youngest Upper Carboniferous rocks (the Coal Measures) being eroded from the central areas and left only on the eastern and western flanks, divided from the limestone country of the central core by a series of high moorlands which mark the Millstone Grit outcrop. Farther north, however, where the upland arch narrows to the 'waist' of the Central Pennines, erosion has not

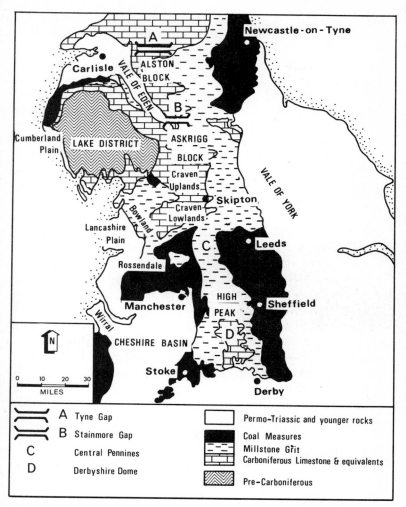

Fig. 25. The main geological regions of northern England and the Pennine moorlands.

succeeded in uncovering the Mountain Limestone so that the high moorland country between Huddersfield and Oldham is entirely a product of the Millstone Grit.

At Skipton the geology changes, for the structure of the Northern Pennines is less simple than that of the south. The arch of the Central Pennines, which becomes progressively assymmetrical northwards, is suddenly broken along its western margin by the Craven and Dent faults. This break at the crest of the arch has caused an uplift of the eastern flanks of the Pennines, the highest points of which now occur nearer to the western margins, not far removed from the fault scarps which face west over Ribblesdale and the Vale of Eden respectively. The uplift has caused erosion to uncover the Mountain Limestone along this steep, western flank and has restricted the outcrop of the newer formations of the Carboniferous, the Millstone Grit and the Coal Measures, to the long, eastward-dipping slopes of the eastern flank. In a few places outliers of the Millstone Grit have survived in patches above the limestone to give the Ingleborough district an unusual type of landscape with its high, isolated mountains rising abruptly from a tableland consisting largely of limestone.

A change in the character of the Carboniferous rocks also takes place in the Skipton district. To the south the Millstone Grit dominates most of the Southern and Central Pennines, the rocks ranging from shales through fine sandstones to coarse sandstones or pebble conglomerates. These were formed in large deltas created by former rivers flowing from the north, which have now totally disappeared, leaving their deposits to be inundated by the lagoon silts and swamp peats of Coal Measure times, and to be incorporated in the later up-arching of the Pennines. To the north of Skipton this type of Millstone Grit outcrop is narrow and ill-defined, and the landscape is now dominated by the Yoredale rocks. A typical Yoredale rock succession (or cyclothem), as seen in the Wensleydale area, would be as follows (oldest at the bottom):

Thin coal band, sometimes absent
Sandstone (Grit), often current bedded (see below)
Shales, often with marine fossils
Limestones, with marine fossils

The succession thus represents a gradual transition in time through the extensive marine invasions (from which the Carboniferous Limestone resulted), and the coastal deltaic environment, which we have seen as characterizing the geography of Millstone Grit times, to the final Coal Measure swamps.

The Central Pennines

The area between the High Peak in the south and Wharfedale in the north may be described as the Central Pennines. The sombre colouring of the moorlands and the dark gritstone walls and buildings contrast greatly with the bare white crags, the emerald-green fields and the light-coloured walls and farmsteads of the limestone areas of the Peak District farther south (see Chapter 11).

The Millstone Grit varies greatly in thickness, being some 6,000 feet thick between Bradford and Burnley, but there is a marked thinning to the west and to the east. Bands of shales which are incorporated within the grit layers have been worn down to form lower ground, thus leaving the gritstone upstanding as precipitous scarps, such as Stanage Edge near Sheffield or the Roaches in north Staffordshire. The use of the stone for traditional grinding methods is perpetuated in the name Millstone Edge in south-west Yorkshire.

In these cliffs the characteristics of the gritstone are readily seen; it occurs in thick beds of coarse material, often with that oblique arrangement of the layers that is known as current bedding. There is no doubt that the deposits were laid down in shallow water under the action of powerful currents, for in some places conglomerates or pebble beds occur among the more normal grits, and it is obvious that pebbles measuring in some cases several inches across could not have been transported by any gentle stream. The grit bands themselves are often of very coarse texture, and have more than once been mistaken for granite, notably by Charlotte Brontë in her description of the moorland scenery in *Jane Eyre*. But this mistake is perhaps to be understood when it is realized that the most abundant minerals making up the grit are quartz and feldspar, both dominant constituents of granite. It is likely that

Plate 20. *A Millstone Grit scarp near Baslow, Derbyshire. This is Gardom's Edge, formed from the Chatsworth Grit; the effect of weathering on the jointing and the bedding planes can be clearly seen.*

they were derived from some granite mass which was undergoing denudation while the grit was being laid down, and that the former river or rivers which brought the material were supplied with debris largely from this source. Six main grit belts are present, including the Kinderscout Grit, but all show changes in thickness as they are traced laterally.

The great bare scarps of grit often form nearly vertical cliffs where the rocks have broken away along the strong joint planes and these are used as training grounds by rock climbers (Plate 20).

Weathering has often carved the rock into blocks of curious shapes by attacking the mass along any planes of weakness and producing fantastic forms which are familiar on many Yorkshire moors. Perhaps the best known are the Cow and Calf on Ilkley Moor. In many places large masses have been quite detached from their bases to form 'rocking stones'.

The steep gritstone scarps, however, only consist of grit near their summits, for the grit bed is mostly resting on shales. The escarpments, therefore, exhibit on a smaller scale the characters already described in the case of the Cotswolds. Along the base of the grit there is frequently a line of springs, and slips of rock down the scarp face are frequent. Alport Castles, in Derbyshire, a magnificent gritstone pile, is the largest landslip in Britain. In fact the scarps are notably weak, owing to the well-developed joint planes in the grit, along which the beds tend to fall away. It is believed that many of the landslips may be of some antiquity, resulting in part from the excessive steepening of the shale slopes during the closing stages of the Ice Age, when the higher parts of the shale surface were rapidly lowered each summer by 'mud flows' produced during the melting of the frozen surface.

Around the northern end of the Mountain Limestone outcrop in Derbyshire the outcrop of the Edale Shales is particularly responsible for landslips. Overlooking the village of Castleton stands the imposing crag of Mam Tor with its Iron Age hill fort. The hill is formed from alternating beds of sandstone and shale which are constantly collapsing owing to the instability of the underlying Edale Shales. So unstable is the cliff face that locally the hill is known as Shivering Mountain. Occasionally landslips have blocked valleys and formed deep lakes; Mickledon Pond, some two miles from Langsett, has been formed in this way.

Above the scarps, the grit outcrops form broad plateaux and long, gentle dip slopes, for in most areas within the Pennines the rocks are inclined at low angles. These tracts of simple relief, mostly over 1,000 feet high and in many places over 1,500 feet, are generally wide open expanses of bleak, uncultivated heaths and moorlands, crossed by rough walls of dark grit. The grit plateaux are in part sandy and dry, in part wet and peaty where rainfall is heavy and drainage is poor, or where a covering of boulder clay conceals the porous grit. The drier areas are occupied by heather moors, with much gorse, bracken, bilberry, and quite a variety of plants, but the peaty tracts, often known as 'mosses', are mostly covered with cotton-grass (*Eriophorum*), the soft white tufts of which have led to the frequent use of the name 'feather-bed' moss in several parts of the central Pennines. Dr T. W. Woodhead analysed the distribution of the names 'moor' and 'moss' and found that in areas above 1,200 feet, where the rainfall is from fifty-six to sixty-two inches, there are many mosses and few moors, but that at altitudes from 700 to 1,200 feet with a rainfall of up to forty-two inches, moors are many times more frequent than mosses. Heather moors are thus more frequently found on well-drained uplands of moderate elevation, as on Ilkley Moor, and also near Huddersfield on Crosland Moor, while cotton-grass moors are very extensive in south-west Yorkshire.

On the barren plateaux there is but a scanty population, for only small areas are now cultivated. These plateaux were, however, extensively settled in prehistoric times during a period of warmer climate, judging by the number of stone circles, cairns and early field boundaries which exist. The dark gritstone farmsteads, which normally occur in sheltered hollows, are generally concerned with sheep-grazing and on the more exposed parts many derelict cottages tell of failure in these harsh conditions. In view of the few habitations on the high Pennine moorlands the countryside has been put to other uses; the earliest of these was the creation of artificial reservoirs by damming the headwaters of the larger rivers, particularly those of the Derwent and Etherow systems. More recently the appearance of high television transmitters at Holme Moss and Emley Moor and the telecommunications tower near Wincle has given variety to the inhospitable moorlands. One of Britain's first 'mountain motorways' is the latest sign of Man's impact on the Central Pennines, for the M62 links the industrial complexes of the Lancashire and Yorkshire coalfields. Apart from the woodlands around the Derwent reservoirs this is generally a treeless area with practically no trees growing above 1,250 feet, due partly to exposure and partly to pollution, since the moors are close to the industrial areas of the West Riding and east

Lancashire. Even the heather is dying out in the smokiest regions, leaving the more vigorous crowberry to expand into irregular masses which form islands in the dreary stretches of dark brown peat. Peat deposits are very widespread, and are often eight feet in thickness, but they are being cut through by streams or eroded by wind in many places, and little peat is being formed at present in most areas. It has been suggested that the obstructed drainage resulting from the irregular deposits left by ice was perhaps responsible for some peat formation, and certainly the deepest parts of many Pennine peats were formed as long ago as Neolithic times, when birch trees contributed to their growth, but the upper parts are largely made of cotton-grass, *Sphagnum* or bog-moss, heather and other moorland plants.

Between Rochdale and Huddersfield, where the Pennines are at their narrowest, the upfold is asymmetrical, with the dips to the west being steeper than those to the east. Accordingly the scarps on the west are more crowded and less elevated, the wide stretches of gently dipping grit to the east having retarded denudation on that side. One of the most prominent of these escarpments is Blackstone Edge, about ten miles west of Huddersfield, which is formed by the Kinderscout Grit; Defoe amusingly spoke of this ridge as the English Andes. Between this and the western edge of the Yorkshire coalfield at Huddersfield there is a succession of similar crag-capped edges, running parallel to Blackstone Edge, and dissected by deep dales.

Most of the familiar dales of west Yorkshire are valleys cut by east-flowing streams in these uplands: Wharfedale, Airedale and Calderdale all represent narrow strips of lowland extending far up into the high moors, and these have served as important routeways across the Pennines from earliest times. The rivers and their tributaries, which occupy still narrower valleys or 'cloughs', have dissected the plateau surface, but between the valleys there are generally great areas of the plateau which are little affected by them. Along the valley sides the grits form lines of crags, while the shale slopes are often covered by fallen blocks and by landslipped masses.

In the dales waterfalls are very numerous, for the alternate bedding of the hard grit and the softer shale furnishes conditions admirably suited for their development (Fig. 26a). Examples are so numerous that it is perhaps unnecessary to name more than two; the Lumb Falls near Hebden Bridge are caused by the Kinderscout Grit, which also gives rise to the fall in Marsden Clough, at Holmbridge near Huddersfield.

A rather different type of waterfall is sometimes produced where a fault line running across the valley brings grit against shale. If the shale is on the downstream side of the fault it becomes rapidly worn down to a lower level than the grit, so that the fault plane is excavated and forms a wall over which the water falls; if, as is frequently the case, the fault plane is nearly vertical, the grit bed may not be undercut for some time, and there is thus less tendency for the fall to recede and to produce a gorge. An admirable example of such a fall is that known locally as the Dolly Folly waterfall, near Meltham, where a little stream falls over a fault face showing over thirty feet of massive grit (the Huddersfield White Rock). This is brought against soft black shales, which are being rapidly eroded and which are seen in front of the fall dipping steeply downstream (Fig. 26b).

Because of the differences in rock hardness the Pennine streams frequently have 'stepped' gradients, as the water rushes rapidly down to lower levels, now plunging over waterfalls, then for some way flowing more slowly until it reaches the fall formed by another hard band. The contrast between the steep and irregular gradients of such streams and the gently inclined beds of larger rivers like those of the Midland plain needs no emphasis; here in the higher regions most streams are actively cutting down their beds, producing deep valleys which are only slowly widened (as by landslips) and thus making little effect on the upland surfaces, while in the lowlands the sluggish rivers cannot cut vertically since they are little above sea-level for most of their courses, but are able to widen their valleys by their shifting meanders and so reduce further the level of considerable areas.

The gorges of the uplands show many signs of the erosive activity of the rivers. In many torrents great masses of boulders lying on the river banks testify to the power of the waters to move heavy loads, especially at times of flood. On the surfaces of many

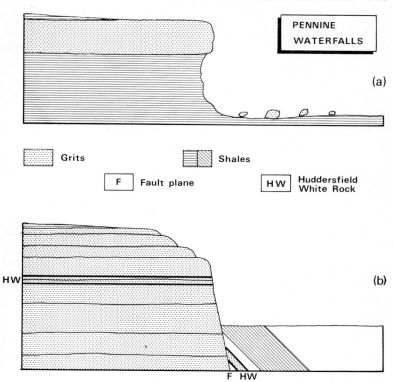

Fig. 26. *Pennine waterfalls:* (a) *A waterfall formed by a hard grit band.* (b) *Dolly Folly waterfall, formed as a result of faulting.*

sandstone beds are cylindrical pots or pot-holes cut by boulders when swirled by the torrents. The growth of a series of such pot-holes rapidly leads to the destruction of a hard band and represents one way in which the lowering of the stream-bed is brought about.

In the Yorkshire dales trees are more frequent than on the up-lands, birch and oak at times being abundant. The wider dales are more populated than the uplands, and there is some cultivation. Where they are crossed by old routes there are grey stone towns: the ancient narrow bridges have determined the location of many of these. The local grits have been used in the buildings, with flagstones sometimes for roofing, their dull colours fitting into the austerity of the view. The grits have been used in building the churches as well as the cottages and walls; their use appears to have begun at a very early date, and it is surprising to find that they were also carried into east Yorkshire and Lincolnshire for church-building in pre-Norman times.

With the development of the woollen industry, at first merely a supplement to sheep raising, the population of the dales steadily increased and the expansion of home weaving led to the building of stone cottages with large 'weavers' windows'. The introduction of water power caused the industry to prosper still further in these valleys beside the rapid streams, which later supplied the lime-free water needed for other stages in textile manufacture. But the greatest expansion of the industry took the bulk of the population on to the lower ground farther to the east, where the dales open out on to the Coal Measures. Here, in the West Riding, a series of old towns has grown to industrial eminence, though among their newer and less attractive scenic developments they have occasionally retained some traces of their earlier character.

The elevation of the coalfield is much below that of the moorlands, but running across it are several prominent edges due to hard sandstone bands occurring among the more usual grey shales of the Coal Measures. Much of the lower ground in the coalfield was formerly wooded, and trees are common even now, especially in the hedgerows which take the place of stone walls in this country. Stone was much less freely used in building than was the case in the moorland dales, and old half-timbered buildings of the sixteenth century still survive in some of the less industrialized towns.

A little distance to the west of the central Pennines a somewhat similar upland tract, known as the Forest of Rossendale, juts out into Lancashire. This region is situated between the south Lancashire coalfield, around Rochdale and Bolton, and the Burnley coalfield. It is thus of anticlinal structure, the coal-bearing rocks dipping away from it very much as they do from the grit areas of the Pennines. The plateau is made up of Millstone Grit which is closely comparable with that already described, the rocks in this case being gently folded. In the middle of the area, in the region including Haslingden and Ramsbottom, the beds are practically horizontal, and form a terraced landscape, each step corresponding to the outcrop of a grit bed, while above it a sloping shelf marks the presence of softer beds. Near the borders of the upland area both towards Rochdale and Accrington the dip increases and the rocks give rise to a succession of scarps and dip slopes. The whole region, however, is cut by numerous faults and its structure is much more complicated than would appear from this brief account. It is deeply trenched by narrow valleys in which hard grits give rise to water-falls, many of which were utilized for power in the early days of cotton spinning.

The Askrigg Block

North of Skipton the character of the Pennines changes, for the narrow belt of gritstone moors flanked by the coalfield conurbations is replaced by a wide stretch of moorland, more than thirty miles across, rising to heights of over 2,000 feet. This area, sometimes known as the Askrigg Block, owes its different character to a change in the nature of the Carboniferous rocks. The Mountain Limestone reappears to the north of the Craven fault but the Mill-stone Grit, already becoming increasingly shaley, decreases in thickness to remain only in scattered patches. The upper layers of the Mountain Limestone Series also change for they begin to show a complex interbedding. This is made up of a rhythmic sequence or cyclothem (repeated up to eleven times) of limestone, shale, grit and sometimes coal, which, as we have seen, is known as the Yoredale beds. The type of scenery produced by these rocks is best found near Ingleborough. Capping the summit of Ingleborough itself is a small area of Millstone Grit and the mountain may thus be considered as an outlier of the once continuous gritstone layer which formerly covered the whole of the Pennine arch. Other remnants of this same bed can be seen on the summits of Whern-side and Pen-y-Ghent at heights of more than 2,000 feet.

Below the gritstone capping, the slopes of Ingleborough, Whern-side and Pen-y-Ghent are carved from Yoredale rocks with the terraces and scars formed from the several limestones. Streams which rise near the summit have cut channels across the Yoredales before they cross on to the Mountain Limestone plinth from which these summits rise. It has already been noted in Chapter 11 how the drainage thence disappears underground. Wherever boulder clay does not mask the rocks their different characteristics are emphasized by the vegetation. The acid soils of the grits and shales carry only heather moors or mosses overlying the peat beds, the dark colours of which contrast with the bright green turf of the limestone areas. The two alternate in strips where the Yoredale beds have produced a terraced land-surface.

Any viewpoint will show that the isolated 2,000-foot peaks of the Ingleborough district must have been carved from a formerly continuous surface of Millstone Grit. This period of denudation resulted in the formation of a lower platform at about 1,300 feet, above which these peaks rise and into which the dales have subsequently been eroded. Such a nearly level surface may have been produced by marine action, that is by wave advance and the cutting of an almost level platform just below sea-level, or it may have resulted from river erosion, meandering rivers and their tributaries reducing the area practically to base-level. Whichever agency was responsible for the formation of this platform, it involved a very long period of denudation at a time when the present plateau surface was near to sea-level. Many plateaux owe their characters to the uplifting of old surfaces, reflecting distant periods of denudation whose actual date is often only approximately known. Reference is later made to the uplifted surfaces in Wales, but it may be noticed that in each case the raising of the surface has given rivers the opportunity to cut deep valleys and so has led to its dissection. On most of these erosion surfaces, moreover, it is possible to recognize residuals which escaped major erosion and which now stand out more or less conspicuously above the ancient plain, as do Ingleborough and Whernside. Such residuals are sometimes spoken of as 'monadnocks'. It is possible that this Pennine platform was carved in late Tertiary times, though some geologists believe that it is much older.

The Alston Block

North of the Askrigg Block the Pennine highlands narrow once again at Stainmore where a Roman road climbs gently up the eastern dip slope from Barnard Castle before dropping steeply down the western slopes into Westmorland and the Vale of Eden. To the north of Stainmore the high moorlands increase in elevation as we follow the Pennine Way up Teesdale past Caldron Snout, while away to the west Cross Fell, at 2,930 feet the highest point of the Pennines, rises above the featureless moorland plateau.

This area is often referred to as the Alston Block, similar in its geology to the Askrigg Block, for the Yoredale rocks again predominate over the less extensive Millstone Grit. Structurally, it is of great simplicity (Figs. 27 and 29); on the north, west and south it is cut off from the adjoining areas by faults of some magnitude, so that it may be thought of as a structural unit consisting of an almost rectangular block of gently tilted and generally unfolded rocks. It is the northernmost portion of the Pennines, and in some respects resembles the part which forms the moorlands of Yorkshire. Stretching from the borders of that county to the Tyne valley, the higher parts of this desolate upland are chiefly occupied by peaty moorlands and by bleak pastures. Reaching more than 2,000 feet above sea-level along its western edge, it falls gradually

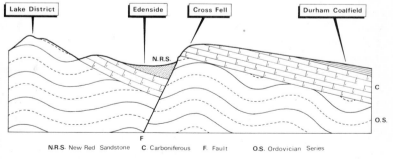

N.R.S. New Red Sandstone   C. Carboniferous   F. Fault   O.S. Ordovician Series

Fig. 27. *The structure of the Alston Block and Edenside.*

eastwards to sink almost imperceptibly into the rolling grasslands which border the Durham coalfield. The general slope of the surface is in approximately the same direction as the dip of the rocks, which are tilted towards the east; the western border, along the line through Cross Fell, forms a great scarp produced by a fault which brings the New Red Sandstone of the Vale of Eden in contact with the older rocks of the moors.

In these uplands the only town of any size is Alston itself, which, although in Cumberland, is situated on the south Tyne in the north-west part of the moors. A dark, stone-built town, it is one of the highest market towns in England. Rising above it on all sides are wide tracts of nearly level moorland. Deeply trenched as this is by many valleys, there remain many expanses of undissected plateau. This upland represents an old erosion surface comparable with and perhaps of similar age to that already referred to around Ingleborough and Whernside, a surface planed down by denudation and subsequently uplifted to its present height. Standing out from this undulating plateau are detached mountain masses, Cross Fell, Great Dun Fell, Mickle Fell and Cold Fell, which probably represent greatly reduced residuals of even older erosion surfaces, now almost totally destroyed.

The heights of the plateau and its residuals are to a large extent unrelated to the nature of the rocks composing them, and harder and softer rocks alike build parts of the surface. Many of the streams flowing across the plateau occupy narrow, gorge-like valleys indicating that dissection of the area has not progressed very far. Most of these valleys may be looked on as having been formed by consequent streams, flowing with the inclination of the plateau towards the east. Tributaries of the Wear and Upper Tees are the most important of these. As in the Yorkshire dales, settlements and cultivation penetrate thus into the moorlands, with many small fields and meadows separated by grey stone walls. Grey farm buildings, roofed by heavy slabs, are scattered in these dales, but the increasing demand for water supply in the nearby industrial complex of the Durham coalfield has led to the creation of reservoirs in some of the Weardale and Teesdale tributary valleys. Where Mountain Limestone appears in Upper Teesdale a number of very rare plant species appear in a community unique in Britain, and in the same area occurs the coarsely granular 'sugar' limestone which is also unrepresented elsewhere. In the whole of Britain Sir Dudley Stamp selected this area as '. . . the locality most sacrosanct for conservation, and where scientific results are most likely to be of far-reaching importance'. Yet this is the latest valley to be threatened by reservoir construction. Although not faced with anything like the famous Teesdale flora, we shall see how similar conservation problems are facing the planning authorities in the valleys of central Wales.

Perhaps the greatest interest of the Alston Block lies in its borders. Of these the western edge is by far the most impressive, falling away sharply along a straight line coinciding with the Pennine fault, which determines the great scarp beneath Cross Fell to the low ground of Edenside (Fig. 27). In this low country the New Red Sandstone forms a deep inlet extending southwards from the Carlisle plain between the Lake District and the Northern Pennines, a long tract of pretty, undulating land almost hemmed in by mountains. The steep western slope of the Cross Fell range is drained by swift streams flowing into the Vale of Eden. Many of these have deep, picturesque valleys, wooded in their lower parts; the valley of Geltsdale, north-east of Castle Carrock, is among the best known of these. But they only furrow the scarp face, and the Pennine edge is not crossed by any conspicuous valley between the Tyne Gap on the north and the Stainmore Pass in the south.

Both these breaks in the Pennine watershed, like the Yorkshire dales to the south, have determined the routes of roads and railways from west to east. These valleys were initiated mainly during later Tertiary times, so that the essential characters of the drainage pattern were determined before the Ice Age. But the valleys played a significant part in controlling the direction of ice movement, which in some cases caused important modifications to the pre-glacial landforms and drainage. Such modifications have been worked out in great detail by a number of authors, but it is not proposed to discuss them fully, for they are comparable with the changes to be described in the succeeding chapters on the Lake District and Wales.

Ice Action in the Pennines

The ice sheets which invaded the Pennine areas came mostly from the north and west. In the western areas they gave rise to ice marginal lakes which produced important meltwater channels, and were instrumental in forming Rudyard Gorge in the Leek area of the south-western Pennines, and Cliviger Gorge south-west of Burnley. This western ice was held up by the high land of the moors, some of which may have escaped glaciation during the last ice advance. In the more northern parts the ice movement was partly guided by those valleys which cut through the uplands, particularly by Teesdale; these received ice from the Lake District and from Scotland. Other valleys which have closed heads received little ice from this direction, but ice formed on the high ground at their heads and gave rise to local valley glaciers. During the periods of maximum glaciation large, erratic boulders were scattered on many uplands, giving evidence of the former extent of the ice. Particularly well known are the great perched blocks of dark grit which stand on the limestone platform at Norber near Settle.

It was during the retreat of the ice, however, that some of the most striking of the existing features may have been produced. The shrinkage of the ice uncovered many parts of the uplands and of the smaller valleys. Some authorities have suggested that on the projecting uplands, especially on the Millstone Grit edges, frost shattering and other 'periglacial' processes helped to freshen up the tor-like forms which are commonplace in many Pennine moorlands. It is important to note however, that other authorities are equally emphatic that the gritstone 'kops' or castle-like rocks are simply the products of weathering under climatic conditions similar to those of today. The shrinking valley glaciers, comparable in form with those of modern Norway, built end-moraines whenever their retreat was checked by climatic fluctuations, for much of the transported material was deposited where the glacier ended. If the end of the glacier remained in one position for some time a considerable heap was accumulated, whereas when retreat was more rapid the material was spread less conspicuously over the whole distance of the retreat. In many dales a series of five or six such moraines extends at intervals almost completely across the valley. Many of these morainic heaps formed dams across the valleys after the ice had retreated above them, and for a time held up the drainage, deposits accumulating in a narrow lake until the dam was breached and the lake drained. In the Aire valley there are well-marked moraines between Shipley and Keighley, one extending across the valley at Bingley, where the river has cut a gorge to avoid it on the south-west side. In Wharfedale, Ilkley Church and Leathley Church stand on morainic mounds.

To the east, on the lowlands, the glacial deposits are more striking, and the Vale of York ice sheet built a moraine which provides a ridge crossing the marshy lowlands from the Wolds to the hills of west Yorkshire, which has been of great importance in the history and development of York itself.

13. The Scottish Border Country and the North-Eastern Lowlands

As we leave the Pennines and travel north into Scotland there is little to tell us when the border is crossed, for it is neither a marked physical nor a cultural boundary. Northumberland, Cumberland and the Scottish border counties share the common cultural heritage of a marchland, and this is also reflected in the continuity of the rolling fell country which passes without a major break from the Northumberland Fells and the Cheviot Hills into the Southern Uplands of Scotland. Indeed, the major physical boundary lies well to the south at the Tyne Gap, where rivers have picked out the fault-guided northern perimeter of the Alston Block to furnish a corridor through the 'Backbone of England'. This area lay on the northern frontiers of Roman Britain, with Hadrian's Wall as a testimony to the early strategic importance of the Tyne Gap. As further proof of its marcher function the region has a higher density of castles and fortified buildings than anywhere else in Britain. A glance at a map will illustrate how the towns of Carlisle and Berwick-on-Tweed were key garrisons, commanding the only viable crossing-places of the rivers Eden and Tweed respectively.

The remarkable affinity between the landforms on both sides of the border is characterized not only by the similarity of land use but also in the use of local topographical names across the administrative boundary; the broad, undulating plateaux with their culminating summits ('laws'), the narrow and often deeply incised dales which rise up to steep amphitheatre-heads ('hopes'), supplied by gorge-like tributary valleys ('cleughs'), while the small lakes of the plateau have already become 'loughs'.

North of the Tyne the limestones, which helped to give to the Alston Block some of its most spectacular scenery, begin to diminish in thickness as they are gradually interleaved with beds of sandstone and shale. Eventually, to the east of the Cheviot, the limestones become 'cementstones' (clayey limestones) interbedded with gritstones, and this combination is important because of its influence on the drainage pattern of the area. With the exception of the Cheviot itself these Northumberland hills are considerably lower than the Alston Block of the Northern Pennines, partly a response to the absence of Mountain Limestone. In this part of Northumberland the Yoredale beds of the Carboniferous are much thicker than in the Alston Block. Nevertheless, the Cheviot Hills are at the centre of a dome-like structure and here the Yoredale rocks have been considerably eroded from the centre of this dome. Although the Yoredale rocks are very much older they have a similar arrangement to those of the Weald, where alternating hard and soft rocks are exposed in an eroded dome to give a landscape of concentric inward-facing scarps and vales. A similar picture prevails in Northumberland, where the parallel scarps are formed from the gritstones, especially the Fell Sandstones, and the vales from the softer cementstones or shales (Fig. 28).

The most marked line of hills sweeps in a broad curve southwards from Berwick. It is formed from an easterly tilted outcrop of Fell Sandstones and rises gradually in elevation as it is traced southwards past the attractive stone town and castle of Alnwick, through the picturesque village of Rothbury to the forested plateau of Simonside (1,409 feet). This line of hills shuts off almost completely the Cheviot Hills from the Northumberland coast and is cut through by only two rivers, the Aln and the Coquet. Both of these, which rise on the slopes of the Cheviots and gather many tributaries from the strike vales of the softer rocks, succeed in breaking through the scarp of the Fell Sandstones where it is affected by faults. In addition to the fortifications at these two river gaps, forts and castles are commonplace along the whole length of the ridge, from Chillingham Castle (with its herd of ancient white cattle) in the north, to Chipchase Castle (based on an old peel tower) in the south.

The scarp formed from the sandstones faces westwards towards the centre of the former dome, and as the older and softer

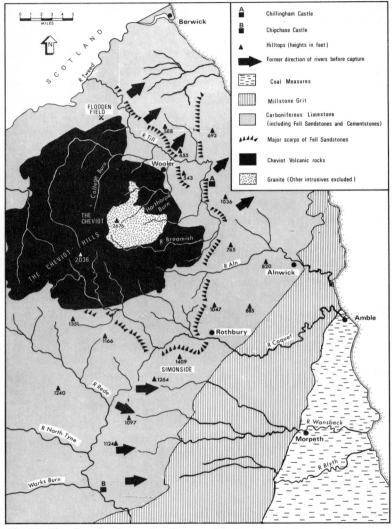

Fig. 28. *The effect of geology on landforms and drainage in north-east England.*

shales and cementstones emerge from beneath this hard bed so the land falls away as a broad concentric vale which can be traced northwards to the valley of the Tweed. Apart from the coastal route this vale, curving along the eastern flanks of the Cheviot and drained mainly by the north-flowing river Till, has been an important border routeway from earliest times; battle sites along its length (including Flodden) testify to its significance in the stormy history of the border.

One of the most interesting features in the more recent geological history of the area is the way in which the former radial drainage of the dome has been severely modified by the presence of softer rocks. The consequent streams would formerly have drained eastwards or south-eastwards down the dipslope of the Upper Carboniferous rocks, as do the Aln and Coquet today. Owing to the comparative ease with which the tributary streams could erode the softer rocks which now make the concentric vale, however, the drainage of the northernmost and southernmost rivers has been captured by piratical subsequent or 'strike' streams. In the north the river Till has led off the College burn, Harthope burn and the river Breamish to join the Tweed; in the south the North Tyne has captured the river Rede which originally formed the headwaters of the Wansbeck.

The Cheviots

Passing westwards beyond the Carboniferous rocks towards the border, the great dome of Cheviot creates the highest and wildest tract in Northumberland, nearly twenty miles across, round which the scarps just described form irregularly concentric belts. The Cheviots reach a height of over 2,600 feet but they include no impressive mountains: rounded peat-covered slopes are typical of much of the area, such rocky crags as occur being inconspicuous. The Cheviot itself, the highest part of the area, is one of the least remarkable of mountains, for the smooth summit is covered with thick peat, but it affords splendid views of the north of England. In places peat is still growing, but as in the Pennines many of the deposits are undergoing denudation, and they are deeply furrowed.

In this lonely brown upland, rivers have carved some picturesque, deep gorges, but speaking generally it shows few of those more impressive features which might be expected on mountainland built up of igneous rocks. For the Cheviot region is best described as a deeply dissected volcano.

Since Carboniferous rocks rest on the Cheviot lavas and dip away from them, it follows that the volcano is of pre-Carboniferous date. It has been established that the lavas date from the earliest part of the Devonian period, at which time they were poured out to form a great mountain on an ancient land-surface. Into this tract other igneous rocks were also intruded, giving rise to the large mass of pink granite which occurs in the middle of the upland. The resulting landforms have suffered greatly from the denudation which began during Devonian times and has been renewed at different periods; finally the whole was overridden by ice in glacial times, leading to still more smoothing of the surface and to the burial of much of it under a cover of ice-borne deposits. In short, the Cheviot Hills, whatever the attraction of their peaty summits or heather-covered slopes, are disappointing as examples of topography owing to the presence of igneous rocks. Although some of the rocks have been used for rough walling, surprisingly few local rocks have been used in building, the isolated shepherds' dwellings being built from freestones carried into the area.

The glacial features of the Cheviots and the country around them are of particular interest. Dry valleys, frequently cutting across spurs between neighbouring stream courses, are very numerous, and represent meltwater channels cut during glacial retreat. On the lower ground are deposits of boulder clay and gravel comparable with those in the Tyne Gap. Kettle-holed moraines supporting small lakes, such as that south of Cornhill, are typically developed.

The Whin Sill

It has been seen how most of the high plateaux and scarps of Northumberland are formed from sedimentary rocks of Carboniferous age. Of quite different origin is the ridge along which Hadrian's Wall extends north of the Tyne at Haltwhistle (Plate 21). Here a

Plate 21. *Hadrian's Wall sited on an outcrop of the Whin Sill near Housesteads, Northumberland.*

bold, craggy scarp faces the north, and from its summit the rolling country may be seen stretching away to the Scottish borders. This scarp, however, is not formed by a sandstone or by any sedimentary rock; it results from a body of igneous rock known as the Whin Sill (Fig. 29). The rock forming this mass varies somewhat, but it is mostly dark blue-grey in colour (as seen on a fresh and unweathered specimen). It is very similar to the basalt of the Derbyshire lavas in composition and in many other respects, save that it is normally of coarser texture: the rock is known as a dolerite. It was formerly regarded as a lava flow of Carboniferous age like the basalt, since it always occurs between Carboniferous rocks. More detailed study, however, and especially the mapping of its distribution, have revealed that the Whin does not always occur between the same two Carboniferous beds. Thus if it were interpreted as a lava it must be of different age at different places, since it sometimes comes among higher beds in the series, sometimes among lower beds. Under Bamburgh Castle it is seen to cut across successive beds of sandstone. Moreover, in places the sill splits into two or more separate sheets. From the field relations it is therefore clear that this rock cannot have been poured out as a lava flow making a definite bed amongst the other rocks of the Carboniferous series, and it is now known that the material composing the mass was injected into the rocks at some date after they had been formed. In parts it has a bedded appearance, but more frequently it is cut by numerous joints which give rise to a weak columnar structure comparable with that which is familiar in the Giant's Causeway.

This injected or intruded material was molten and in many ways similar to a lava at the time when it found its way from some deeper source along the weaker places between the beds, sometimes leaving one level to find a higher or lower path where it met less resistance. In the Whin rock we have an example of an igneous rock, formed from the consolidation or crystallization of molten material, but which, unlike a volcanic rock, solidified underground and not at the surface. The slower cooling allowed the growth of larger crystals than are to be found in such rocks as basalt, which, having been cooled on the earth's surface, has crystallized more quickly and has a finer texture. It is interesting to note that along the margins of the Whin Sill the rock is sometimes indistinguishable from a basalt, for the reason that the molten material was there rapidly chilled by its contact with the bounding rocks.

The thickness of this intruded sheet or sill naturally shows great variation. In places it exceeds 200 feet, but the average thickness is more nearly 100 feet. This sill occurs over a very extensive area, coming to the surface at places determined by the structure, for although its relations are not exactly those of a bedded rock, its form is comparable with that of the Carboniferous strata among which it occurs. With them it has been arched into wide folds or broken by faults. To some extent, therefore, the Whin Sill behaves as a particularly tough bed among the shales and sandstones where it occurs, and its scenic effects are mainly due to its resistance to erosion. The great scarp already referred to, crowned by the Roman wall on the west of the North Tyne valley, is similar in general character to any escarpment formed by sandstones or limestones, though the dark rock in places forms more rugged crags; frequently, too, the soft shales beneath the dolerite have been scooped out by ice action, and small tarns and mosses are common.

The most northerly occurrence of the Whin Sill is in the Kyloe Hills, whence the outcrop trends south-eastwards to the dark cliffs under Bamburgh Castle (Fig. 29); still farther east its course can be traced in the low and rocky Farne Islands, the existence of which is due to the resistance of the tough dolerite when the softer sedimentary rocks have been worn away from around them by the waves. To the south again, in the columned cliffs of Cullernose Point, near Alnwick, the sill again forms dark vertical cliffs, but at that point it ceases to be a factor in the coast scenery, its outcrop turning inland to form the ridge mentioned above, and extending southwards along the high fault scarp which bounds the Alston Block. It forms the well-known crags in High Cup Nick above Appleby.

In the wide vale above Middleton in Teesdale the Whin Sill is again exposed, forming striking scarps. Six miles above Middleton it gives rise to the well-known waterfall of High Force, the dolerite acting as a hard band in exactly the same way as the gritstones in the Yorkshire dales; the brown, peaty waters of the Tees fall over a

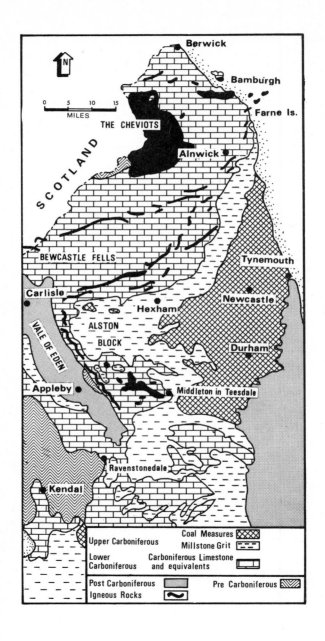

N

0 5 10 15
MILES

Berwick

Bamburgh

Farne Is.

THE CHEVIOTS

Alnwick

S C O T L A N D

BEWCASTLE FELLS

Tynemouth

Carlisle

Newcastle

Hexham

ALSTON

BLOCK

Durham

VALE OF EDEN

Appleby

Middleton in Teesdale

Ravenstonedale

Kendal

Upper Carboniferous

Lower Carboniferous

Post Carboniferous Igneous Rocks

Coal Measures

Millstone Grit

Carboniferous Limestone and equivalents

Pre Carboniferous

vertical face of dark rock to a swirling pool seventy feet below, whence a wooded gorge extends downstream and marks the retreat of the fall. Farther upstream Caldron Snout is another striking fall caused by the resistant Whin Sill in the river's bed, the water cascading down great stairs of black rock.

The Coalfield and the Coast

The coalfield region of the north-east of England is not treated in detail. It is sufficient to note that in Northumberland steam coals and house coals are most prevalent, whilst in Durham coking and gas coals are the rule. Most of the thicker seams are now exhausted, having supplied the bulk of London's coal since Elizabethan times. The coalfield consists mostly of country of lower altitude than that farther west, but since there are thick sandstone beds among the shales and softer rocks of the Coal Measures, there is great diversity within the coalfield. In particular, a group of sandstones which occur just above the most important series of coal seams is responsible for some of the highest ground in the coalfield. It gives rise to the steep hills in Newcastle itself, while its outcrop extends southwards to Low Fell. In Durham the dips in the Coal Measures (as in the higher ground to the west) are very low, and the sandstones frequently form almost horizontal cappings to even-topped hills. Along the coast these same sandstones are responsible for much variety in the cliff scenery north of Tynemouth.

To the east of the Coal Measure outcrop lies a wide tract of red and brown rocks belonging to the New Red Sandstone, the most important member of this series being the Magnesian Limestone. This rock group forms an almost straight outcrop from Nottingham, where it is weak and inconspicuous, through Knaresborough to the coast, where it extends from Hartlepool almost to South Shields. The Magnesian Limestone is a yellowish-brown dolomite, a limestone with a considerable proportion of magnesium carbonate, but like other limestones it is frequently cavernous owing to

Fig. 29. *The geology of north-east England and the location of the igneous rocks of the Whin Sill and the Cheviot igneous complex.*

the action of the underground waters, while large areas on its surface are streamless.

The Magnesian Limestone forms the coastline of Durham for many miles, giving rise to steep, yellow cliffs which are in marked contrast to those formed by boulder clays and by Coal Measures farther north. In places, as at Marsden, tall stacks have been left standing by wave erosion along vertical joints. The cliffs are also varied by attractive 'denes' cut by the east-flowing streams.

Inland, the Magnesian Limestone forms a sharp scarp facing to the west and crossed by few rivers. It is possible that many of the consequent streams from the Alston Moors formerly continued eastwards across this tract, but if so they have been captured by the middle portion of the Wear, which from Bishop Auckland to Chester-le-Street is evidently a subsequent stream flowing at the foot of the limestone ridge.

There is little doubt that the Wear formerly continued still farther northwards to join the Tyne at Newcastle, for a deep buried channel known as the Team Wash marks its earlier course; this channel is now almost completely filled in with glacial deposits, and as the deepest part is over 140 feet beneath sea-level it must have been cut when the land stood a good deal higher than at present. The Wear left this course owing to the changes brought about during the Ice Age, and it now turns sharply eastwards through the Magnesian Limestone escarpment to Sunderland. This course of the lower Wear is determined by an overflow valley cut when the Wear drainage was dammed up by ice to the north. As in the case of the Pickering lake the deepening of this newly established course led to its persistence after the ice had disappeared and left masses of glacial debris blocking its former course. The deep channel at Ferryhill probably represents a still earlier outflow from the Wear Lake, before the ice had retreated so far north. For much of its lower course the present Wear thus occupies a valley which has been cut entirely since glacial times. Upstream, however, the valley is deeply incised in solid rock and at Durham a sharp meander forms a hairpin gorge from Elvet Bridge to Framwellgate Bridge; the almost completely detached portion of the plateau enclosed within the bend of the river forms the splendid site of the cathedral and castle.

14. The North-Western Lowlands

It is a striking fact that in northern England the rivers which flow
to the east coast are generally twice as long as those which empty
into the Irish Sea from the western slopes of the Pennines. We have
seen in Chapter 12 how the asymmetrical arch of the Pennines has
generally resulted in the highest land being found along the western
flanks of the Central and Northern Pennines, and this usually coincides with the watershed. One important product of the asymmetry
of northern England is the contrast in the width of the lowlands
on either flank of the Pennines; in Yorkshire the land below 1,000
feet stretches some sixty miles between the West Riding and the
North Sea, whereas in Lancashire it is on average less than half this
width. Farther north the contrast between east and west becomes
even more apparent, for the highland mass of the Lake District
squeezes the Lancastrian and Cumbrian coastal lowlands to insigni-
ficance. Not until we reach the Carlisle plain and the Vale of Eden
do we find western lowlands to equate with the broad, fertile Vale
of York and the Durham–Northumberland coastal lowlands.

The asymmetry of northern England is no accident but is a
response to the distribution of the older and younger rocks, which
can be said to correspond roughly to 'highland' and 'lowland'
Britain. A glance at a geological map of Britain will demonstrate
how sedimentary rocks younger than the Coal Measures generally
coincide with the lowlands of southern and eastern England. Conversely, the rocks older than the Coal Measures usually form land
over 1,000 feet in elevation, because the majority are harder, and
therefore more resistant to erosion, than those of south-eastern
England. If we add to this the knowledge that north-western
Britain has experienced several periods of uplift throughout its

geological history it becomes possible to explain the differences in relief between highland and lowland Britain.

Thus the broad lowlands and low rolling hills and scarps which lie to the east of the Pennines are carved mainly from rocks that are younger than the Coal Measures and which extend as far north as Tynemouth. It is significant, therefore, that the only lowlands which occur in Lancashire and Cumberland generally coincide with structural basins of New Red Sandstone (and some Coal Measure rocks) which lie within the highland province of old, hard rocks. Because they are found within the highland zone, however, these softer rocks have often survived only by virtue of down-faulting, and therefore as the boundary faults are reached the land frequently rises steeply on all sides as one crosses to the harder rocks. Nowhere is this more clearly demonstrated than in the Vale of Eden.

The Vale of Eden and the Carlisle Plain

In the previous chapter it was seen how the Pennine Fault cut off the western flank of the Mountain Limestone of the Alston Block to form the massive escarpment of Cross Fell. While the fault has allowed the uplift of the hard rocks to the east it has also caused the preservation of the New Red Sandstone which has been down-faulted to the west and subsequently picked out by rivers to form the Vale of Eden. There is a striking contrast in scenery and land use between the high, barren and drab Pennine moorlands and the fertile, chequerboard field-pattern of Edenside. One is immediately reminded of the similar landforms and scenery of the Vale of Clwyd in North Wales, when viewed from the Clwydian Hills, and it is no surprise to discover that the geology and structure are almost identical.

On the flanks of the Cross Fell fault scarp a break in the steep slope is marked by a line of foothills, at an elevation of some 1,400 feet. This step in the escarpment is caused by a narrow belt of very old, hard rocks, similar to those of the nearby Lake District, which form the 'basement' of northern England and which have been exposed by the denudation of the downfaulted New Red Sandstones (Fig. 27).

The composition of the various New Red Sandstones, termed the Penrith Sandstone, the St Bees Sandstone and the Kirklinton Sandstone, as they are traced upwards, is of very great interest, for it enables us to reconstruct the landscape of the region when they were being formed during Triassic and Permian times. Some of the oldest beds, which were used by the Romans to aid in the construction of Hadrian's Wall and have been quarried near Kirkby Stephen and Appleby for local buildings, are termed 'brockrams'. A brockram is a coarse, angular scree or breccia, consisting mainly of fragments of Mountain Limestone in a matrix of red desert sandstone. One can visualize a narrow desert valley, similar to the modern Death Valley of California, surrounded by towering uplands of older rocks. The screes on these mountain slopes would be swept out by occasional torrents to form gravel fans that eventually intermingled on the lower ground, and were subsequently buried by dune sands. The bright red Penrith Sandstone, from which much of the attractive market town of Penrith is built, lies not far from the Mountain Limestone which here skirts the eastern flank of the Lake District and forms the western edges of the Vale of Eden. It is important to note, however, that the contact is not a faulted one, as on the eastern side of the vale. Thus the river Petteril, which rises near to Penrith and flows north to Carlisle, has succeeded in picking out the western boundary of the red sandstone and, in working along it as a strike stream, has left a marked west-facing sandstone scarp, which acts as a watershed between the Petteril and Eden rivers that flow parallel to each other from Penrith to Carlisle.

At Carlisle the sandstone ridge disappears, but not before providing a bluff upon which the old fortified city was founded. Although most of the city is brick-built many of the older civic and ecclesiastic buildings are constructed of New Red Sandstone.

In the Carlisle plain the red marls and shales with gypsum beds lie directly on the sandstones, and cast our minds back to the red Keuper Marls of the Midlands, to which they are thought to be related. From Carlisle these softer rocks extend westwards past Wigton to the coast at Silloth, but rarely do they outcrop at the surface owing to their thick cover of glacial drift. Only around

Great Orton, where the glacial deposits thin out, do solid rocks form a small tableland which stands above the plain. This is fashioned from the youngest sedimentary rocks of north-western England, dark bluish shales and limestones of Lower Liassic age, which suggests that the desert conditions had given way to a marine environment heralding the onset of Jurassic times. We can only pause to wonder at the enormous amount of erosion that must have occurred in order to destroy these newer rocks, which must once have linked this small outlier with the more extensive Liassic outcrop in east Yorkshire.

One of the most important factors influencing the scenery of the Vale of Eden and the Carlisle plain has been the thick glacial deposits which were brought from Scotland and the Lake District during the Ice Age. The effect of the underlying rocks on the texture and composition of these glacial drifts has played an important part in the different types of land use in the region. Generally the lowland soils are formed from the New Red Sandstone drift and are sandy, friable and relatively easy to work, so that ploughland occupies almost half of the vale area. Where boulder clay creates less tractable soils, especially in the areas of underlying marls, and on the wide stretches of alluvium of the valley floors, cattle thrive on the lush grassland. On the flanks of the vale, however, where the harder rocks outcrop, the soils are poorer and more acid, partly because the glacial drifts are thinner or less extensive, and partly because of the heavier rainfall. One passes, therefore, fairly rapidly from the rich, farming country of the lowlands, with their coppice woodlands and hedgerows, to the poorer grasslands and drab treeless moorlands of the Pennines and the Lake District where sheep grazing and stone walls predominate. The outcrop of Mountain Limestone which fringes the western margin of Edenside has had little effect on the soils and scenery of the area, for ice movement from the Lake District has spread a stony and clayey mantle over the slopes, giving a soil more suitable for pasture than for crops, and forming a landscape fundamentally different from that of the red soils of the lowland farms.

A final word is needed on the landforms created by the glacial deposits. They can be divided roughly into two types: the clay features and the sand and gravel features. Boulder clay either gives rise to featureless ground, sometimes marshy, or under certain conditions it can result in ridges and hollows which parallel the direction of former ice movement. In the Vale of Eden some of the ridges break up into 'drumlins', smooth hillocks, typically oval in plan, and with the 'upstream' ends steeper and blunter than the others. These have obviously been fashioned by moving ice and play an important part in the local scenery. Hedgerows frequently follow the crest of their 'whale-back' ridges like dorsal fins, while roads either wind crazily around their flanks or mount steeply over their tops, having previously plummeted into the intervening hollows. Sands and gravels result from the melting of the ice sheet, and the country covered by such deposits may be more diversified and sharper in feature than that covered by boulder clay. The Brampton area shows these steep-sided features very clearly, where ridges rise for more than 100 feet above the general level and form a belt some eight miles long and several miles wide, probably marking temporary halting places during the ice retreat. Such 'kames' frequently enclose small lakes or 'kettle-holes', representing places where masses of ice were stranded in the deposits to leave an enclosed hollow when the buried ice at length melted; in some cases such lakes have subsequently become peat bogs.

The land bordering the Solway is low and flat, with marine clays stretching inland for several miles. Located upon these are large peat bogs such as those of Bowness and Wedholme Flow, but more important are the marine clays which are covered by a turf with short, springy grass, that finds a ready market for lawns and sports grounds all over Britain.

The Cumberland Coastlands and Morecambe Bay

South of Maryport the Lakeland fells draw near to the coast as the outcrop of the New Red Sandstone runs offshore. The coastal plain narrows and the landscape changes abruptly as we pass southwards across the Maryport fault on to the Cumberland coalfield. The usual colliery clutter and regimented rows of brick houses are familiar sights on any coalfield, although the majority of the pits are now closed in this isolated outlier of industrial Britain. The Carboniferous rocks also contain rich pockets of very pure hematite iron ore which has stimulated an important iron and steel industry centred on Workington. A contrast is at once seen with the other large iron and steel town of the north-west, for Barrow-in-Furness is located as a seaport near other iron-ore supplies rather than on a coalfield. At the southern boundary of the Cumberland coalfield, where the St Bees Triassic Sandstone forms a prominent coastal headland, Whitehaven shows the choice of building stone available at such a location; while the harbour was built from Coal Measure sandstone, the churches were constructed with St Bees Sandstone. Nevertheless, the chemical factory, based on the anhydrite of the Triassic rocks, is now one of the most dominant features of the town. These western industrial districts thus give to the mountain-land a seaward fringe of smoke and tip-heaps.

Only between St Bees Head and the Duddon estuary is the coastal plain given over to the farming landscape which is usually associated with New Red Sandstone country. Even here, however, in the shadow of Lakeland, the nuclear reactors of Calder Hall show a twentieth-century variation on the traditional juxtaposition of agricultural and industrial landscapes, which typifies the Cumberland coastal plain, as well as south Lancashire and Midland England, wherever the Coal Measures project through the New Red Sandstone.

· Near Millom, where the majestic mountain of Black Combe brings the Lakeland fells almost to the sea, the narrow down-faulted outcrop of New Red Sandstone is cut off from its southern counterpart in Furness by the Duddon estuary, which marks the Lancashire border. Here, with its wide expanse of honey-coloured sands where flocks of sheep graze on the lush green turf of the coastal marshes, is an area which has inspired numerous writers, including Wordsworth. For the combination of high mountains and moorlands mirrored in estuaries and bays of the sea is rarely found south of the Scottish border. We shall see how a combination of similar features in North Wales has given to parts of that coast too a scenic beauty unsurpassed in southern Britain. As in North Wales, the presence of old, hard rocks, which here make up the Lake District, bringing high summits, steep slopes and a different colouring into the coastal scene, adds something which the New Red Sandstone fails to do in its entire length between the Solway and the Dee, despite the splendid cliffs at St Bees Head.

Morecambe Bay repeats on a larger scale but in a more subdued tone the picture so vividly portrayed at Duddon Sands. The small, isolated patches of New Red Sandstone which occur in Furness and the Cartmel peninsula must all be continuous beneath the waters of the bay. We see here, then, how denudation has lowered the structural basin of Triassic sandstones to such an extent that the post-glacial rise of sea-level has inundated all but a few vestiges of their former outcrop. This has brought the sea into contact with the older rocks which ring the sandstone basin, and since these are composed of rocks with varying degrees of resistance to erosion the coast of Morecambe Bay is fretted with creeks and tidal inlets. Many of its bays and headlands are carved from Mountain Limestone and the white rocks, grassy hollows and dark green woods 'cling like a garland about the hem of the capricious sea' (Edwin Waugh). Whether the capricious sea will be excluded from the bay in the future by a man-made barrage, as at Portmadoc (see p. 266), remains to be seen.

The Lancashire Lowlands

South of Morecambe Bay the plain becomes wider and is crossed by the two largest rivers of these lowlands, the Ribble and the Mersey. The increasing width of the coastal plain is in response to the reappearance of the Triassic rocks south of Heysham, where the sudden rise in elevation of the country east of Garstang coincides with the faulted boundary of the Triassic sandstones and the harder Carboniferous rocks which build the Forest of Bowland to the east. From Preston northwards, canal, railway and motorway keep closely to this marked physiographic boundary (Fig. 30).

Although New Red Sandstone makes up much of the floor of the Lancashire lowlands the solid rocks rarely emerge from beneath their thick covering of glacial and post-glacial deposits. The landscape in western Lancashire, therefore, depends to a larger extent

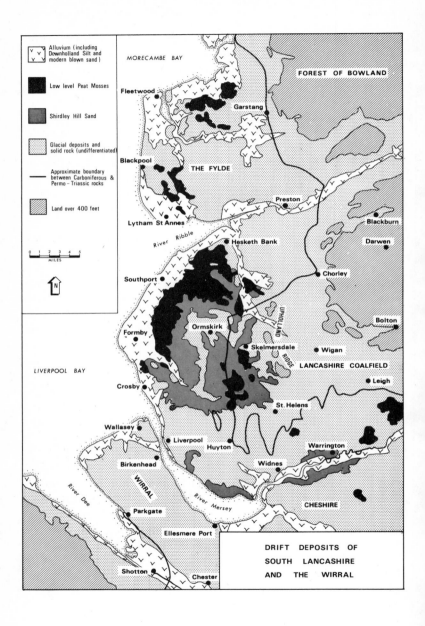

DRIFT DEPOSITS OF SOUTH LANCASHIRE AND THE WIRRAL

on the detailed contrasts in soil and drainage characteristics of this drift cover than on differences within the New Red Sandstone sequence (Fig. 30).

The thick drift cover continues eastwards to the slopes of the Pennines, and despite the slight increase in elevation of the country to the east of Skelmersdale and St Helens, it would be difficult to judge where one had passed from Triassic to Carboniferous rocks, were it not for the dramatic change in the man-made features of the landscape. For there are two Lancashire landscapes here on the lowlands: the farmlands of the west and the industrial townscapes of the east, the boundaries corresponding roughly with the junction of the New Red Sandstone and the Coal Measures.

Between the river Lune and the river Ribble the solid rocks play no part in the make-up of the region known as the Fylde, whose name, derived from the Old English '*gefilde*', means the 'plain'. The undulating surface of boulder clay with occasional mounds of glacial sands and gravels is today an ordered landscape of neat fields, well-trimmed hedgerows and carefully enclosed copses and spinneys. It is difficult to realize that this rich farmland, with its green pastures carrying a heavy density of livestock, was a creation only of the last two centuries. Before the eighteenth century this area, and also that of south-west Lancashire, presented a totally different appearance, for it was an area of dreary marshlands, known as mosses, which stood behind the coastal dunes. Apart from the clearance of the natural woodland the most impressive change in west Lancashire has been the transformation of these 'mosses' into the rich pastureland described above.

As the ice sheets melted and left behind their blanket of boulder clay, which forms the coastal cliffs of north Blackpool, so the badly drained hollows soon became infilled with post-glacial peat. Place-names incorporating 'moss' and 'mere' occur frequently in the area today but give little indication of their former meaning except for the abundance of carefully maintained ditches and drains. The rich black soils of the former peaty depressions, however, enable one to distinguish the differences, despite the general grassland cover.

Fig. 30. *The drift deposits of south Lancashire and the Wirral.*

East of the river Wyre, Pilling Moss and Rawcliffe Moss are now completely farmed but vestiges of the former landscape can still be glimpsed at Winmarleigh Moss.

The land which skirts both shores of the Ribble estuary and the entire western coast past Southport to the Mersey has had a different post-glacial history, for the deposits which were formed above the boulder clays have played an important part in the land use of the area. Several thousand years after the withdrawal of the ice sheets, the sea-level, rising throughout the world as the water returned to the oceans, crossed the line of the present coast in western Lancashire. With no solid rock to bar its progress this marine transgression inundated the plain of glacial drift and left a low cliff-line, now much degraded, in the boulder clay a few miles inland. The old shore-line may be traced intermittently from the Ribble to the Dee but is best seen at Hesketh Bank, near Preston, and at Hill House, east of Formby. West of the cliff-line the boulder clay is everywhere mantled with a stiff blue marine clay, known as the Downholland Silt, although in many places this is buried beneath thick peat deposits similar to those already described. Altcar and Halsall Moss, between Liverpool and Southport, are two of the former mosslands now reclaimed. In places covering the Downholland Silt, but more frequently extending several miles inland from this old marine cliff-line, is a deposit referred to as the Shirdley Hill Sand. This represents the dunes associated with the former shore-line, in much the same way as the dunes at Formby, Southport and Lytham St Annes are found along the present coast. Apart from contributing to a zone of soils which are fairly heavily exploited for market gardening because of their tractability in an area of heavy clays, the Shirdley Hill Sands are extensively used in the south Lancashire glass industry because of their refractory properties.

These western coastlands remained very thinly populated until the end of the eighteenth century. The barren sand-dunes of the present coast, the mosslands studded with meres, and the heavy clay soils of the thickly forested glacial drift plain, all contributed to a landscape which was hardly conducive to early settlement. Thus there is virtually no evidence of either prehistoric or Roman

Plate 22. *The resort of Southport built mainly in Victorian times on the coastal dunes of south-west Lancashire after the draining and reclaiming of the hinterland marshes. Beyond the eastern boundary of the town the peaty soils of the former marshes are characterized by their farmlands, their coppiced woodlands and their virtual lack of settlement. In the distance the Upholland ridge, an outcrop of Millstone Grit, divides the industrial landscape of the Lancashire coalfield from the farmlands of the plain.*

occupation in the area. Not until the draining of the mosses do we see the planned Victorian resort of Southport emerge on the coastal dunes (Plate 22), and although Defoe speaks of Liverpool's remarkable expansion during Queen Anne's reign it was not until the Industrial Revolution that this great Mersey port grew rapidly away from its waterfront.

Only near to the Mersey and the Dee estuaries does the New Red

Sandstone emerge from beneath the drift cover to play any significant part in the landscape. The Wirral Peninsula will be treated separately below, but at Liverpool the city has expanded from the blown sand and alluvium of the estuary shore up on to the Keuper Sandstone ridge of Everton and beyond to the Bunter Pebble Bed hills of Anfield and Walton. A thin wedge of Bunter Sandstone helps to create the ridge on which the cathedrals stand and through which the railway has to tunnel on its way to Lime Street station. The famous Olive Mount cutting at Edge Hill is some two miles long and eighty feet deep.

Although the Coal Measures extend westwards almost to Ormskirk and Huyton, they are buried beneath a thick cover of boulder clay, so that it is not until one crosses the Upholland ridge to the north of St Helens that one becomes aware of a different landscape. This ridge, which rises some 500 feet above the plain, is mainly the result of a faulted slice of Millstone Grit emerging through the Coal Measures. From a vantage point such as Billinge Hill we can ponder on the differences in geology and geography which have contributed to the contrasts in landscape which lie before us. To the west are the bright green farmlands, the coppiced woodlands, the black soils of the reclaimed mosses, the dark line of the forested Formby dunes with the glint of the Irish Sea beyond; to the east, the colliery winding gear, the forest of chimneys, the subsidence lakes or 'flashes', the railway tracks, the canals, the spoil heaps, known ironically as the 'Wigan Alps', all scattered between the endless smoke-blackened brick towns of the Lancashire coalfield which stretch almost without a break into the cotton towns of east Lancashire. The presence of a number of good coal seams at the surface made this coalfield one of the first centres of the Industrial Revolution. All the surface workings are now exhausted, and although deep mining continues, the colliery waste is gradually being levelled, the flashes infilled and new housing estates and recreation grounds are replacing the derelict areas. Nevertheless, a recent writer has been prompted to remark how '. . . the pallid fields wear eternally a veneer of industrial grime, while the Irwell and its tributaries are little more than open drains for the waste of all the mills upon their banks' (Millward).

There are barren parts in the St Helens–Widnes areas too, although here the mountains of slag are products of the chemical industry which grew up where coal and Triassic salts and sands were readily at hand.

One of the striking features of lowland Lancashire is that nowhere in the landscape can we evoke a feeling of contact with the remote past. We have seen how the western coastlands, from Morecambe Bay to the Mersey, were unattractive to early settlement owing to the mosslands and poor soils. The earliest heritage that can be distinguished, therefore, relates to the drainage schemes of the eighteenth century, for a few of the windmills, built to drain the water and grind the first crops of corn, have survived in the Fylde. On the coalfield, however, the industrial development has almost obliterated any vestige of the earlier landscape. A few half-timbered houses have survived, as at Hall-i-th'-Wood at Bolton, to remind us that had it not been for the Industrial Revolution, the 'magpie architecture' of the Cheshire plain (Plate 17), which formerly extended across the Mersey into the Lancashire lowlands, would have been commonplace in a region almost devoid of building stone.

The Wirral Peninsula

Beyond the river Mersey, which in places has been artificially deepened and channelled to accommodate the Manchester Ship Canal, the Triassic rocks of Lancashire continue southwards into the Cheshire basin. As in Lancashire the covering of boulder clay is generally thick enough to obscure the solid rock, although in the Wirral peninsula small hills of Triassic sandstone rise above the Plain.

The general level of the peninsula is about 150 feet, and this low platform frequently coincides with the occurrence of the Bunter Pebble Beds. Since these are usually mantled with glacial drift, however, the landscape is one of farmland, not forested or heath-covered hills as at Cannock Chase and Whitmore in Staffordshire. Wherever the Keuper Sandstone occurs the Wirral landscape changes, for this hard rock usually creates sudden ridges, such as

Heswall Hill, Caldy Hill, Thurstaston Hill and Bidston Hill, with their red rocks and gorse-clad slopes. Even more striking is the forested ridge of Storeton Hill which hides the oil terminals and the factories of Merseyside from the farmlands of central Wirral.

The pleasantly wooded landscape of Wirral, where the tiny streams have cut picturesque gorges through the boulder clay into the underlying sandstones, as at Barnston, has attracted an increasing proportion of the Merseyside population during the present century. The countryside is therefore rapidly disappearing beneath vast housing estates which are creeping slowly out from Wallasey and Birkenhead. As a result the half-timbered buildings, which once existed here, have been destroyed and the attractive black and white village of Thornton Hough is only a Victorian copy of a lost heritage. A few sandstone buildings have survived and even a windmill remains in the northern half of the peninsula, but generally suburban brick is paramount.

The unusual shape of the Wirral Peninsula has provoked an interesting hypothesis concerning its evolution (Fig. 30). It has been suggested that the Mersey formerly followed a course from Runcorn to Shotton where it joined the Dee in pre-glacial times. The present narrow and deep estuary of the Mersey is thought to be a trough scooped out by the ice sheet which drove south-eastwards into the Cheshire basin during the Ice Age. Whether ice was responsible for the lower Mersey channel is not certain, although there is a great deal more evidence to support the former course of the Mersey through to the Dee where the old channel now lies deeply buried beneath glacial drift.

There is an interesting contrast between the estuaries of the Mersey and the Dee; whilst the former is a deep waterway lined with docks, oil refineries and large industrial enterprises, the latter is a rapidly silting, marsh-fringed bay, similar in some respects to Morecambe Bay. The little village of Parkgate, with its quay and picturesque 'waterfront', was, after the decline of Chester as a port, an embarkation point for Ireland, but today the tide never reaches it because of the growth of marshes on the Gayton Sands. Silting of the lower Dee has also led to the demise of Chester as a port, and to avoid flooding the river has been canalized from here to Shotton.

The marshlands, however, in contrast with those of Morecambe Bay and the Solway, are rapidly being reclaimed and developed, with the vast Shotton steel works now dominating the landscape. The contrast in shape between the estuaries may account for this difference in character. In the Mersey, tidal flow is constricted by the narrow strait between the sandstone hills of Liverpool and Wallasey, thus assisting channel scour. By contrast, the Dee estuary widens considerably seawards and no such tidal scouring can be expected.

15. The Lake District

The scenery of the Lake District is so well known and has been so eloquently described that there is no need to praise it here. Since Englishmen have overcome their dread of desolate heights and have learned to appreciate mountains, the Lake District has attracted so many writers that many of its place-names have become part of English literature. Wordsworth and the other lakeland poets, Ruskin and more recent writers have made its richness known so that its popularity has become almost a danger. Much of this region of 'lovely rock-scenery, chased with silver waterfalls' is still almost unspoiled, however, except along a few main roads and beside some lakes; it is still impossible to see the greater part of its mountains except on foot, and while these conditions remain at least the worst results of popularity are not likely to be too prominent.

The Lake District is a small and compact area. Within this tract, only about thirty miles across, the craggy mountains, bare fells, deep ravines, impressive falls, and lakes of differing pattern afford examples of scenic features nowhere else to be seen in England. Although the increasing numbers who travel to the Alps and farther afield in search of mountain scenery have the advantage of larger lakes and higher peaks, those who best appreciate the greater mountain masses find their enjoyment of the Lake District enhanced; its mountains are small but it is easier for the ordinary traveller to know them better, while the lakes and cascades show all the essential characters of Swiss scenery, with an added daintiness if with less vivid colouring. Only the glaciers are wanting, and they have not long disappeared.

This area is of high relief with heavy rainfall and generally acid soils, so only small tracts within its valleys are cultivated; where the valleys open out on to the surrounding lowlands there are wider agricultural areas, and here are situated the old market towns, such as Cockermouth, Penrith and Kendal. Within the uplands, apart from the larger tourist centres, there are only small villages and isolated hamlets built of the grey local stones and often tucked away in lonely valleys.

The systematic description of the Lake District, area by area, is beyond the scope of the present chapter. It is proposed rather to describe those features peculiar to its scenery, with references to some examples, and to indicate their origin; to discuss, in the words of Ruskin, 'first what material there was to carve, and then what sort of chisels, and in what workman's hand, were used to produce this large piece of precious chasing and embossed work'. The materials of the Lake District are old, older than those which occupy any considerable tract in the areas already described; they are comparable, however, with the old rocks which underlie the Carboniferous in Cross Fell. But although the rocks are old, the features incised on their surface are new; the valleys have no greater antiquity than those of the Pennines or of the London basin, while the lakes are even younger.

The Influence of Rock Types on Scenery

The old rocks of the Lake District are often hard, but the height of these uplands is not primarily due to this fact. While it is generally true that older rocks are mostly harder than newer ones, there are many exceptions; on the other hand, there are many areas of low ground which are occupied by quite resistant rocks. For example, the rocks of Anglesey are in part older and harder than those of the Lake District, yet that island is almost a monotonous plateau only a few hundred feet above sea-level. The altitude of any area is naturally determined in the first place by the extent to which it is raised by earth movements: the Lake District is believed to have been so raised during mid-Tertiary times (when the pene-plains of the Alston Block and middle Pennines were also elevated). If exposed to the action of denuding agents for a sufficiently long time the softer rocks will be greatly lowered while the harder beds will stand out; to a certain extent this has already occurred, but on the whole the uplift is so recent that soft as well as hard materials are still contributing to the upland surface.

The scenic differences due to the varying materials which build up the Lake District may be understood without going far from such popular centres as Keswick or Ambleside. From the Friar's Crag by Derwentwater, possibly the best-known viewpoint in the area, there may be seen the smooth slopes of Skiddaw to the right, beyond Keswick, while to the left at the head of the lake the crags of Lodore and the mountains of Borrowdale show an irregular 'crinkly' skyline and steep rocky precipices of quite different appearance. A line drawn from near the southern end of Ullswater along the east and south of Derwentwater towards Ennerdale separates off an area of slaty rocks to the north from an area of pre-dominantly volcanic rocks to the south.

The rocks of the northern belt are known as the Skiddaw Slates; they include grits and sandstones and are all of sedimentary origin, representing muddy and sandy sediments laid down in a shallow sea. They have since been upraised and severely crumpled, the squeezing to which they were subjected having converted some of the muddy beds into slates, which split into layers along planes produced by the pressure. But the pressure appears to have been rather different from that which produced the cleavage in the North Wales slates, for the Skiddaw Slates are useless for large-scale industrial purposes. Since they occur in thick, slabby masses they are more suitable for dry-stone walls than for roofing, and thus the Skiddaw Slate country is devoid of the vast slate quarries which so disfigure the Cambrian slate belt of Caernarvonshire. Instead the area has a few small mines following the mineral veins which occur; the copper mines are now closed, the barytes mine at Force Crag is working sporadically, but the graphite of Keswick, although completely worked out, has led to the foundation of the famous pencil works which now thrives on imported raw material. Many of the Skiddaw Slates are by no means soft rocks, but they are very much less resistant than the volcanic rocks which lie to the south of them. Yet Skiddaw reaches an altitude very little below that of Scafell and Helvellyn, in the volcanic tract, and higher than that of

Plate 23. *The Langdale Pikes seen from Elterwater village in the Lake District. These high rugged peaks have been carved from the Borrowdale Volcanic Series and are characterized by steep cliffs beloved by climbers. The thickly wooded valley in the foreground has been eroded in a belt of Ordovician slates and shales and heralds the onset of the lower pleasant park-like country that extends from here towards Windermere.*

many imposing summits there. Undoubtedly the Skiddaw rocks are being much more rapidly lowered by weathering and other eroding agents, and the fact that Skiddaw is so high makes it probable that the uplift of the region took place at no very distant date. In this slate region many small streams have cut deep gorges, leaving narrow ridges and sharp peaks, but no wide expanses of level uplands.

Skiddaw is a mountain with few crags, but its steep slopes, covered with bracken and heather, are none the less attractive. It is interesting, too, as one of the first English mountains known to have been climbed for the sake of the scenery and the thrill. For the travellers of just over two centuries ago seem to have been stirred by this 'stupendous mountain' with its 'chasms and enormous depths in the bowels of the mountains'. Farther east Saddleback (or Blencathra) is a more exciting mountain, for it has some sharp crags. Generally, however, the mountains formed of Skiddaw Slates both east and west of Bassenthwaite are of smooth aspect and have few bare rocky cliffs, giving a type of scenery which is quieter and more restful than that beyond the head of Derwentwater.

The rocks which lie to the south are known as the Borrowdale Volcanic Series (Plate 23). They include thick beds of lava together with other rocks formed from the dust or 'ash' thrown out by the volcanoes, and agglomerates made up of a concrete-like mass of broken fragments produced during the volcanic explosions. Among these volcanic materials some beds of mud are included, subsequently converted into slates by pressure. The well-known green slates from Honister Crag above Buttermere form good roofing material, in contrast to the Skiddaw Slates discussed above. There is practically no evidence as to the position of the actual volcanoes from which the lavas were ejected. It cannot be too strongly emphasized, however, that the form of the present mountains is quite unrelated to the original volcanoes; the higher peaks stand out from the rest of the area because of their particular resistance to the agencies which have worn down neighbouring tracts, but not because they mark the sites of the vents. Generally the more compact lavas and the harder ashes tend to stand out as crags and cliffs while the softer beds are worn down, and the extraordinary diversity of the area and the irregularity of the hill-forms are largely due to the variability of the great pile of volcanic material. The strong joint-planes cutting through thick, tough beds give rise to many steep precipices. Indeed, the attractiveness of the Lake District fells to the rock-climber, as in Snowdonia, is due to the weathering of the hard Lower Palaeozoic volcanic rocks into steep and craggy faces.

The mountains in this belt include Helvellyn, Scafell, Great Gable and the Langdale Pikes. The volcanic belt, however, is not so much to be regarded as an assemblage of great mountains but as an irregular upland traversed by deep valleys which are sometimes so narrow as to make little impression on the upland surface: the mountains are the more conspicuous heights left between the valleys and combes. Thus the most familiar traverse through the volcanic country, by the main road from Ambleside by Thirlmere to Keswick, does not help in an appreciation of the real form of the uplands; from the valley the hills seem of very uneven height.

It is from the summits that the form of the country is best appreciated, and Helvellyn and Scafell afford almost equally fine views of the mountains, separated by the deeply trenched valleys. The Lake District, indeed, is so small that from one viewpoint almost the whole of it can be taken in, though every new viewpoint shows a fresh aspect. From Scafell the valleys are seen to radiate in almost every direction. The twin eminences of Scafell and Scafell Pike, joined by the ridge of Mickledore, illustrate the weathering of the massive volcanic rocks. Looking over the precipitous northern face the view embraces the bare slopes of Great Gable and a great expanse of bleak, rocky country with the upland lake of Styhead Tarn, while beyond Derwentwater are the gentler contours of Skiddaw and its neighbours. Great Gable with its cliffs and pinnacles of thickly bedded rock is another fine viewpoint, since it rises sharply at the head of the deep valley of Wasdale. Here the lower ground, occupied by alluvium and by the more uneven glacial deposits, is sufficiently fertile for cultivation, and an irregular patchwork of fields follows the valley floor, but as the slopes steepen upwards, bare crags and screes bring to an end all attempt at cultivation. The view from Honister Crag along the straight and steep-sided valley occupied by Buttermere and Crummock is similar in many respects, although here patches of woodland by the lakes tend to soften the landscape.

The bleak and rather desolate tract of the Borrowdale Volcanic rocks can be traced southwards as far as Ambleside, where it merges into the lower and pleasant park-like country that extends by the shores of Windermere. The highest hills in this tract do not reach 2,000 feet, and their smoother outlines recall the hills of the Skiddaw belt. The long lake of Windermere with its small bays and

Plate 24. *A corner of Hawkshead village in the Lake District. Local slate has been used in a variety of ways; in addition to its frequent use as a roofing material it is here shown as a wall cladding and as treads in the outside staircase.*

bordering stretches of woodland lies between hills of great charm, but only in the distant mountains beyond its northern end is there a hint of wildness and grandeur. The rocks forming this southern part of the Lake District are somewhat like those of the Skiddaw belt, but whereas the latter are older than the volcanic rocks, those of Windermere are newer: these rocks, known as the Upper Slates, also include a varied series of grits and limestones, and form a great series over 13,000 feet in thickness. One of the limestones, known as the Coniston Limestone, has an influence on the scenery out of all proportion to its width of outcrop. It occurs as a narrow band running north-eastwards from the Duddon valley through Coniston, Tarn Hows and Waterhead to beyond Troutbeck, and its interest lies in its effect on the vegetation; in a region of acid soils the white limestone supports lime-loving plants and an altogether lusher and greener ground-cover than elsewhere. It is not a strong topographic feature but it marks an important boundary between the rugged volcanic country of the central Lake District and the more gentle relief of the Silurian rocks to the south.

Resting partly on the volcanic rocks and partly on the Coniston Limestone these Silurian rocks dip to the south and form hills which generally present steeper faces to the north. They are mostly shales, mudstones and sandstones, many of which split easily into slabs about an inch thick. Thus they are termed 'flags', after their common use as paving-stones. Some of the mudstones have been turned into slates by pressure and these are worked profitably for roofing slates, especially at Kirkby-in-Furness. A further use of slates in the Lake District is illustrated in the picturesque village of Hawkshead where slate-hung walls of so-called 'Westmorland blue' (though quarried in Lancashire) blend with the grey stone cottages, some of which have whitewashed elevations (Plate 24).

A discerning analysis of the relation between Lakeland landscape and its domestic architecture is that of Wordsworth who noted that the cottages

. . . of rough unhewn stone, are roofed with slates which were rudely taken from the quarry before the present art of cutting them was understood, and are therefore rough and uneven in their surface, so that both the coverings and sides of the houses have furnished places of rest for the seeds of lichens, mosses and flowers. Hence buildings which in their very form call to mind the presence of nature, do thus, clothed in part with a vegetable garb, appear to be received into the bosom of the living principle of things as it acts and exists among the woods and fields . . .

And such woods and fields characterize the low smooth hills of the southern Lake District where the Silurian rocks have generally produced a better soil than have the Borrowdale Volcanics farther north.

The materials out of which the core of the Lake District has been carved thus consist of ancient rocks belonging to the Ordovician and Silurian, which fall into three groups giving rise to three well-defined areas. These old rocks have been involved in intense earth pressure, which converted the muddy sediments into slates, and impressed on the rocks a complicated structure. Into them, moreover, were intruded igneous rocks, including the granites of Shap, Skiddaw and Eskdale, and some of these form conspicuous hills; the famous gabbro intrusion of Carrock Fell stands out clearly above the Skiddaw Slates which adjoin it on east and south. Of the granites the most widely used as a building stone is the red granite of Shap which is exported to all parts of Britain. On the other flank of the Lake District, between Wasdale and Bootle Fell, a much larger intrusive mass, the Eskdale granite, forms the lower fell country of the south-west. It seems curious that, although we think of granite as a hard rock, the Eskdale granite weathers relatively easily, rotting down to a coarse sand without producing good blocks for building stone. Because of this there is a dearth of dry-stone walls, with hedges being rather more frequent than is usual in the Lake District. Similarly, the Ennerdale granophyre, farther north, although helping to produce the Ennerdale Fells, has weathered into lower rounded foothills from which rise the rugged volcanic masses of High Stile, Pillar and Steeple; the contrast in both shape and elevation is here most marked.

All these intrusions occurred before Carboniferous times; the rocks of the area had been raised to form part of a great range of mountains which were worn down and then submerged beneath the Carboniferous sea. Carboniferous rocks were laid down on the worn surface, almost horizontal beds thus coming to rest on the highly disturbed strata of Ordovician and Silurian age. These Carboniferous rocks now form an almost continuous ring around the older rocks of the Lake District. South-eastwards they pass from Kendal into those limestone areas already described around Settle, while south-westwards they give rise to the beautiful scenery bordering the estuaries near Dalton-in-Furness and Grange-over-Sands. Here the limestone country is marked by long white scars and gentle grass-covered dip slopes, like those of the Warton Crag above Carnforth.

All around this ring, the Carboniferous rocks dip outwards, and their scarps face the higher uplands of the Lake District. It is obvious that these beds formerly extended farther over the uplands,

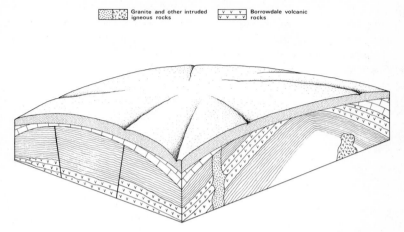

Fig. 31. *The Lake District, showing the initiation of the radial drainage on a domed surface of newer rocks.*

and since an extension inwards of the present dips would carry them well over the tops of the highest mountains, it is highly probable that the limestones were laid down over the whole region, the central portion having been removed after the strata were domed upwards.

The Structure of the Lake District

The relation of the Carboniferous rocks to those on which they rest is different from anything described in the previous chapters, where we have mostly dealt with beds laid down in regular series, each layer resting on that immediately older and the whole forming a continuous and conformable series; the Carboniferous rocks in the Lake District are, however, unconformable to the older rocks, resting in different places on different beds. The complication is made still greater owing to the fact that the New Red Sandstone and probably other newer rocks were laid down over the whole area after the

Carboniferous rocks had been arched upwards and the higher part had been removed, so that the New Red Sandstone in its turn came to rest unconformably, in some places on the Carboniferous, in others on still older beds (Fig. 31). Later the whole was forced up into a dome and the newer rocks have been worn away in the centre (Fig. 32).

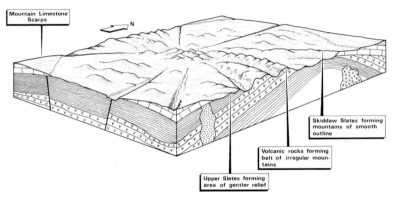

Fig. 32. *The Lake District, showing very diagrammatically the influence of the various rock types on the scenery.*

These somewhat complex arrangements of the rocks have been described briefly in order to facilitate discussion of the origin of the most striking feature of the relief of the Lake District, namely, the radial arrangement of the drainage lines. This peculiar feature was recognized and described by Wordsworth in his *Topographical Description of the Country of the Lakes in the North of England.* Standing on Scafell it is at once apparent that several valleys radiate from the centre; within two or three miles of this mountain seven important valleys extend outwards towards the north, south and west of the Lake District. These include the great Windermere valley (running to the south), the Duddon valley, Eskdale (towards the south-west), Wasdale, Ennerdale (to the west), the Buttermere–Crummock valley, and the Derwent valley (to the north).

The explanation of these radial drainage lines depends on the essentially domed structure of the area. Bearing in mind that the newest rocks known to have covered the area were at one stage arched up into a low dome, it may be regarded as quite probable that the river system was initiated on this surface, rivers flowing outwards in nearly every direction from the centre of the dome more or less as the rain streams outwards from the top of an umbrella. We may suppose that Scafell is not far from the original centre of the dome. Of course the simple domed structure, illustrated diagrammatically in Figure 31, has now almost completely disappeared, the newer rocks having been removed from the greater part of the region as the rivers have cut down: the rivers therefore flow on older rocks of complex structure whose arrangement generally bears no close relationship to the courses followed by the rivers, but on the fringes of the area the rivers pass on to the newer rocks. These changes are shown in Figure 32, where rivers occupying positions closely comparable with those in Figure 31 are shown occupying deep valleys in the uplands formed by the old rocks. The drainage remains radial although the conditions which controlled the initiation have been destroyed as the rivers cut down more deeply.

Such a drainage system, determined by the structure of one group of rocks but later impressed on an underlying group, often of greater complexity of structure, is known as a superimposed system. Not all such superimposed systems are radial, and it is in the combination of these two features that the Lakeland drainage is most interesting. East of Scafell the valleys are not strictly radial in their arrangement, and it may perhaps be worth noting that the uplift was not quite a simple dome: thus there are the Thirlmere valley, Ullswater and Haweswater running northwards from a parting which runs roughly eastwards from Scafell to Shap. This deviation from a truly radial arrangement is, however, only of minor importance, and the accompanying diagrams illustrate a simple radial system. The rivers were established long before the

Ice Age, for the final uplifting of the dome is supposed to have occurred in mid-Tertiary times, when the peneplains of the Pennine regions were raised and when the final folding of the Wealden anticline and the London basin led to the initiation of the drainage systems of south-eastern England.

The Scenic Effects of the Ice Age

The changes in scenery resulting from the Ice Age are of enormous importance. The Lakeland mountain group was an area on which ice formed early during the Ice Age and where it lingered long, while during the time of the maximum extent of the ice it was covered beneath ice sheets of vast thickness. For the most part the ice moved outwards from the centre of the area, thus following the directions of the main valleys, and the most impressive changes have resulted from the scooping out of material along these tracts. In this district and in North Wales, just before the middle of the nineteenth century, the former existence of glaciers in Britain was demonstrated, the similarity of the valleys and their construction to those associated with modern glaciers being so striking as to leave no doubt concerning their interpretation. For long afterwards the character of the ice which left the deposits of boulder clay on the plains of England was uncertain, but the glaciation of these mountain valleys has never been seriously questioned since that time. All the main valleys have many features in common, and description of a few examples will serve to illustrate the essential characters of all.

The valley of Buttermere and Crummock running north-west-wards from Great Gable is sufficiently well known to form a suitable introduction. Viewed from Honister Crag or from any high point near its head, this valley is seen to extend almost in a straight line; the hills which flank it rise with surprising abruptness from the borders of its lakes, or from its floor in the rather small areas where lakes are absent. From the crags at its head the valley falls steeply to a low, green tract which is traversed by the streams entering the head of Buttermere. Beyond this lake stretches the longer water of Crummock and just visible in the distance, almost in the same line, is Loweswater. Though the hills rise so steeply from the valley, they are for the most part less craggy than those of Honister, for the valley is cut in Skiddaw Slates, and the uplands have not the rugged bleakness typical of the volcanic tract. The valley is thus a great trench, open from end to end, the south-western side being especially straight. In many river valleys in high country every turn reveals a new view, every projecting spur hides the next stretch of the valley, but in Buttermere the whole length of the valley may be taken in at one view, and there are no projecting spurs running out from the valley sides. Other good examples of glacial troughs can be seen at Great Langdale and Ennerdale, the upper slopes of the former valley being steep enough to form the vertical cliffs of Gimmer Crag, beloved by climbers, beneath the summits of the Langdale Pikes. These features are essentially the result of glaciation, for while the greater part of the excavation of the trough was done by the river before the Ice Age, the movement of the glacier smoothed out its pattern, and gave it that wide U-shaped cross-section and that simplicity of line which are its most striking characteristics.

The tributary valleys are very different in character, for they are less conspicuous, the streams tumbling rapidly from the uplands and giving rise to numerous picturesque falls in their sharp descent. Some streams, like that which descends from Bleaberry Tarn on the south-west of Buttermere, have cut only an insignificant notch in the steep valley side, and the tributary valleys may be said to 'hang' above the main valley; this is a further result of the work of the ice, which greatly deepened the trench, leaving the tributary valleys high above its floor. On the upland the tributaries are more or less normal streams in shallow valleys, but when they reach the valley they cascade down its side to the new level of the main river, as in Sourmilk Gill, as the foaming stream from Bleaberry Tarn is so expressively named. The streams are cutting new beds more in accord with the over-deepened valleys which they enter, but so little time, in geological terms, has elapsed since the ice disappeared that little progress has been made in most cases.

On the other hand, the high waterfall of Scale Force, made by a stream which descends rapidly to Crummock Water, is situated in a dark gorge, showing that this stream has done more in adjusting itself to the changed conditions. The waterfall occurs where the stream crosses a hard, igneous rock overlying softer Skiddaw Slates, and therefore is rather similar to the more usual English waterfalls, but it may be noticed that most of the cascades made by the streams which dash from their hanging valleys into deep-cut glacial trenches do not depend mainly, if at all, on the existence of hard beds, the falls here being due to steps produced independently of stream erosion.

It is at once apparent that the form of both Buttermere and Crummock is due primarily to their occurrence in a valley; in this they resemble the other large lakes of the region, Windermere, Thirlmere, Wastwater and Derwentwater, which coincide with the lines of main drainage already described. Primarily they are flooded valleys, owing partly to the over-deepening of the valleys by ice excavation, and partly to the blocking of their exits by glacial deposits. The close relation of Buttermere and Crummock is obvious; they are at practically the same level and it scarcely needs to be pointed out that the flat green area which separates them is but the delta formed by deposits carried into the once continuous lake by the stream which comes down the east side of the valley from above Buttermere. Gradually this delta must have grown outwards from the lake margin until, almost reaching the other side, it finally cut the lake into two, forcing the outflow from the upper lake to the extreme border of the valley where it flows between the deltaic flat and the steep slopes of solid rock. Such alluvial tracts, both those at the head of the lakes (for example beneath Honister) and those formed by streams entering farther down the valley, are naturally of great importance in lakeland, introducing small, level, fertile areas in an otherwise bleak and uncultivated tract: in the central part of the Lake District only about two per cent of the land is cultivated, and this is mainly represented by the alluvial flats.

The lakes of the Derwent valley are similar in many respects. Situated in a trough where the ice rapidly deepened the valley on leaving the volcanic rocks of Borrowdale, the hills bounding it take generally the smoother forms characteristic of the Skiddaw Slates.

The tributary streams hang above the main valley and the famous falls of Lodore (disappointing though many find them in dry weather) mark the rapid descent of one such tributary. At the northern end of Bassenthwaite are great stretches of glacial deposits, especially between the lake and Cockermouth. At Keswick the growth of the great delta formed by the Greta on one side of the valley and Newlands Beck on the other has led to the separation of Derwentwater from Bassenthwaite; it is interesting to see how in this instance the presence of the opposite streams has forced the outflow from Derwentwater to find its way to the lower lake between the two deltaic flats, and not along the margin, as in Buttermere, while the abundant material brought down by these streams has extended the delta downstream for a mile beyond their mouths. It is obvious that the continuation of this process must eventually lead to the complete silting up of Bassenthwaite.

Lakes are indeed only temporary features in a landscape, and their abundance in parts of Britain, mostly a result of the Ice Age, emphasizes the short time that has since elapsed. Ultimately most of these lakes must disappear, either by becoming filled up with alluvium or by the cutting of an outlet which allows the drainage of the waters. We have already noticed how pro-glacial lakes of the closing stages of the Ice Age were drained by the latter method in other parts of England; the silting up of the lakes has also been completed in some cases, notably in Kentmere valley, north-west of Kendal, where two lakes, each rather over a mile in length, have been silted up; the upper lake, dammed by a moraine, caught most of the sediment brought down the valley, and is filled by alluvium, but the lower lake, into which the clear water then passed, is partly filled by diatomite, formed from the siliceous remains of lowly vegetable organisms which lived in the lake. Many of the wide, flat stretches in the Eden valley may also be silted-up lakes, while Rosthwaite in Borrowdale, the Naddle valley and St John's Vale (north of Thirlmere) are other examples.

The long ribbon of Windermere shows many features similar to those of Bassenthwaite, but the lower elevation and the gentler slopes of the surrounding hills gives rise to a softer landscape. Near the southern end of the lake are great banks of drift, and these glacial materials, by choking up the lower part of the valley, have been responsible for the formation of the lake, the waters of which now escape seawards, not along the old valley to the south but by the Backbarrow gorge cut through the ridge near Newby Bridge into the valley adjoining it on the west. This gap thus represents another meltwater channel which has become a permanent outlet.

Interesting details of scenic development may be found in all the greater lakes, and possibly enough has been said regarding their major features. The lakes dealt with are all situated on one or other of the slate groups, and most of the larger lakes occupy basins excavated in these less resistant groups. A few, however, notably Wastwater, Thirlmere and Haweswater, are placed on the volcanic series, their steep, rugged sides giving a wilder grandeur to their surroundings. Possibly Wastwater has the finest situation of all the lakes, the straightness of its south-eastern side and the hanging valleys opposite being most impressive. Wastwater is one of the best examples of a glacially over-deepened valley, its surface being about 200 feet above sea-level while its deepest part is over fifty feet below sea-level. At the head of Wasdale is a considerable expanse of alluvial country, but its sides are bare, and apart from small deltas along its length, the rest of the upper valley is bleak. The south-eastern side is famous for its screes, composed of angular blocks broken by frost and other weathering agents from the cliffs above; the loose material extends in fan-shaped masses high into the notches of the precipice, in places almost to the top. The surface of the scree slopes steeply, its slope being determined by the angle of the rest of the blocks. The scree slopes are mostly bare right down to the lake-side, where the material extends below water-level. Although Wastwater is glacially over-deepened, and would there-fore be a lake if all superficial deposits were removed, the water-level is raised by a dam of glacial material at its lower end, near Wasdale village.

The effect of ice action on the character of the scenery is by no means confined to the neighbourhood of the larger lakes. Some fine glaciated valleys are not occupied by lakes; the neighbouring valleys of Long Sleddale, Kentmere and Troutbeck (the southerly Trout-beck, between Windermere and Ambleside), are long steep-sided troughs; the ice appears to have excavated these valleys some 200 or 300 feet deeper than they were in pre-glacial times. Kentmere was then a winding valley and in converting it into its canal-like form the ice cut across its projecting spurs, whose truncated ends now stand out in the great crags of Froswick, Ill Bell, Rainsborrow and Raven Crag.

In many parts of the Lake District, there are great semi-circular basins high on the hillsides, marking the sites of small glaciers which occurred during the final stages of glaciation. They are known as corries or cirques. These corrie-glaciers lingered longest on the shady north-eastern slopes, and corries are extremely numerous in the areas formed by Skiddaw Slates and the volcanic series. Above Buttermere, Burtness Combe and the basin holding Bleaberry Tarn are good examples, while under the Langdale Pikes and in Mardale (Blea Water and Small Water) are many others. The famous Striding Edge of Helvellyn is bounded on the north-east by the great precipice of the corrie which holds Red Tarn. Striding Edge itself is an example of an arete, where two neighbouring corries have encroached headwards sufficiently to destroy the smooth pre-glacial ridge and replace it by a sharp, jagged knife-edge.

Many of these hollows enclose small lakes, the water frequently held up by a dam made by the terminal moraine of the former glacier. Bowscale Tarn, on the southern side of the Caldew valley near Carrock Fell, and Red Tarn on the east of Helvellyn, are typical and fairly accessible examples of such corrie lakes. Great numbers of small tarns are also found in parts of the upland tracts, where they occupy hollows formed in some cases by the excavating action of the ice, in others by the damming of shallow valleys by the moraines. The outflow for some tarns is over such a morainic dam, but for others, even when such a moraine is present, the outflow channel has been cut through solid rock; this must have presumably afforded a lower exit from the lake than the newly heaped-up moraine, just as in the case of the outflow of such large lakes as Windermere. Levers Water, reached by an easy route in about a mile and a half from Coniston, illustrates such a tarn.

16. North Wales

In some respects the scenery of North Wales resembles that of the Lake District, save that its greater area holds wider tracts of desolate upland, its valleys fewer lakes and its longer coastline, which includes Anglesey, has many beautiful bays and rugged cliffs. Moreover, the solid grey castles of Conway, Caernarvon and Harlech, guarding the entrances into Snowdonia, show something of the influence of history on the scenery.

Snowdonia

A beginning may best be made in Snowdonia itself, among the highest mountains of Wales. Here the mountains have many features in common with those of the Lake District, for they include many which are chiefly built up of volcanic rocks, products of volcanoes which were active at approximately the same time as those of that area. Lavas were poured out, frequently on the sea-floor, to form thick beds of hard rock, while ashes from the volcanoes also gave rise to considerable deposits. The actual summit of Snowdon itself is made by such ashes, laid down on the floor of the Ordovician sea, and fossils of marine shells may be collected in this material on the mountain-top. But much of Snowdon is formed by lavas, and by intruded igneous rocks. The structure is complicated, but in general the various beds form a syncline or downfold so that it is clear that Snowdon stands out from its surroundings only because the material which formerly extended around it (and, especially to the north and south, far above it) has been worn away (Fig. 33). As in the case of the Lakeland rocks the actual volcanoes are unknown, but it is probable that the Snowdon volcanic rocks were poured out from a number of vents at no great distance.

From the Snowdon summit the craggy country formed by the volcanic rocks may be seen stretching north-eastwards through the Glyders and the Carneddau, and south-westwards into the Lleyn peninsula of western Caernarvonshire. This country has been profoundly modified by glaciation, and the deep valleys cut it up into blocks and greatly influence the accessibility of the different parts. Snowdon lies between three such main valleys (Welsh: *nant* = valley), the Llanberis Pass (Plate 25), the Nant Gwynant and the Cwellyn valley, and may thus be approached easily from several

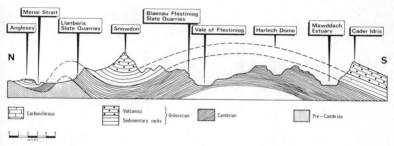

Fig. 33. *The structure of North Wales.*

different directions. The least exciting ascent, from Llanberis (or by the railway which takes the same direction), shows something of the steep precipices. There are many small lakes on Snowdon (Welsh: *llyn*; pl. *llynau*), most of them held up by morainic dams. From the Llanberis track the deep indigo pool of Llyn Du'rArddu is seen cradled in a dark cwm, backed by one of the steepest rock-climbing faces in Wales.

The Llanberis Pass has so many points of similarity to those of the Lake District that the effects of glaciation need not be again described. In its lower part are the twin lakes of Llyn Peris and Llyn Padarn, separated by the delta of the Afon Arddu (Welsh: *afon* = river) which rushes heavily laden down the slopes of Snowdon; higher up, the valley is narrower and more rocky as the road climbs sharply to the top of the pass (Plate 25).

Plate 25. *The Llanberis Pass, North Wales, looking north-westwards across Anglesey. The glacial overdeepening of the valley is clearly seen and the valley lakes of Padarn and Peris, divided by a post-glacial delta, occupy the middle distance. The rugged terrain of the foreground is carved from the Snowdonian Volcanic Series, whilst the Cambrian slate belt beyond is picked out by the Dinorwic slate quarries.*

Beyond the Glyders the Nant Ffrancon runs nearly parallel to that of Llanberis, and exhibits still more clearly the features of a glaciated valley; Nant Ffrancon played an important part in the early study of the effects of land ice. Extending inland from Bethesda to Capel Curig, this pass makes a sharp bend near the lower end of Llyn Ogwen. The northern part of the valley, seen from that viewpoint, stretches wide and open in an almost straight line to

Bethesda, the floor an irregular patchwork of fields, with a few scattered farmsteads on the marshy flat which occupies the former lake floor. The sides rise to sharp bare crags so that the valley section is typically U-shaped, and the spurs are truncated. Several tributary streams rush down the left side, breaking through the small moraines as they flow from cwms which hang considerably above the over-deepened floor of Nant Ffrancon. On the other side of the valley there are no impressive cwms, for as is usually the case in the northern hemisphere, small glaciers lingered longest on the hillsides facing north-east.

This straight and simple section of the Nant Ffrancon valley ends at Ogwen Cottage in a sharp step, the valley floor having that exaggeratedly stepped profile which is produced by ice action. Travelling up the pass the valley here seems to end suddenly against a high rocky cliff; down this the river falls in a most beautiful cascade, escaping from Llyn Ogwen, situated almost immediately above the fall, by a shallow gorge, on the flanks of which great masses of ice-smoothed rock are dotted with perched blocks. As one climbs from the lower wide valley over this rocky step it is surprising to find the valley once more opening out into the smoother tract which encloses Llyn Ogwen and which extends on to Capel Curig. It is believed that the drainage from the Llyn Ogwen valley formerly flowed eastwards to the river Llugwy but the pre-glacial watershed was first breached by an early ice sheet and subsequently lowered by later ice advances.

Llyn Ogwen is obviously a valley-lake, taking its form from the shape of the valley it occupies. It is joined by streams coming down the steep slopes of the Glyders to the south, their courses showing another step. The stream which joins the Ogwen at Ogwen Cottage rushes down this step in a series of torrents which in some places have already succeeded in cutting shallow gorges. Following this stream up from the main road, it may be traced over a huge moraine which stands on a rock bar behind which lies a gloomy cwm holding Llyn Idwal (Plate 26). As one climbs above the col between Glyder Fawr and Y Garn it is possible to see the point at which the glacially smoothed rocks give way to the frost-shattered summit rocks of these two peaks. The lack of glacial smoothing suggests

Plate 26. *Cwm Idwal, North Wales. The glacially eroded hollow has been scooped out of a synclinal structure which occupies the centre of the picture. The cliffs of the Devil's Kitchen can be seen immediately beyond the lake, Llyn Idwal, above which is the glacially smoothed col which descends to the Llanberis Pass. The lake is held in by a crescentic moraine which can be seen perched on the crest of a rocky outcrop in the right foreground. Ogwen Cottage is the small settlement in the central foreground.*

that summits over 3,000 feet protruded as 'nunataks' above the ice cap and were subject only to severe frost action and other 'periglacial' processes during the later phases of the Ice Age.

The cliffs at the head of the lake exhibit a most spectacular synclinal structure, formed in volcanic lavas and ashes. The cliff is seamed with deep gullies, the most striking of which is the Devil's Kitchen (Twll Ddu) down which tumbles a waterfall. Immediately east of the fall the climbers' haunt of the Idwal Slabs obviously forms one flank of the syncline, whilst just north of these, four morainic ridges running obliquely up the hillside mark successive downwasting stages of the former Idwal glacier.

To the north of Snowdon, along a belt which runs through Llanberis to near Bethesda, is the most important slate-quarrying region in Britain, from where the best slates in the world have long been obtained. The more accessible parts are thus scarred by great purple-grey quarries, and disfigured by heaps of refuse from the workings (Plate 25). Slate is used in this belt for many purposes for which stone (or even wood) is employed elsewhere. The area is otherwise attractive, lacking the rugged aspect of the volcanic tract, and more often wooded in the lower parts. These slates, belonging to the Cambrian, are older than the rocks of the volcanic series and derive their character from the intense pressures to which they have been subjected.

The great expanse of mountain country, which lies to the east of Nant Ffrancon and extends thence to the Conway valley, is much less accessible, for this area of some eight miles by ten miles is cut by no important valley and for that reason is crossed by no road. Owing to their remoteness its mountains (including the little-known Carnedd Llewelyn, less than eighty feet lower than Snowdon) are rarely visited but offer a variety of glacial phenomena including Llyn Cowlyd which occupies a deep glacial breach north-east of Capel Curig. Most of the Carneddau lakes have been utilized for hydro-electric power or for local water supplies. But this mountain tract is better known near the coast, where, although it is little over 1,000 feet high, it rises sharply above the sea at Llanfairfechan, Penmaenmawr and Conway. Here the plutonic and volcanic rocks produce craggy hills of considerable beauty; owing to their lower elevation they have a richer vegetation than the rocky heights of Snowdonia, and the brilliant colours of Conway Mountain in late summer, the deep purple heather mingled with golden-yellow gorse, are not to be surpassed in any upland area. Unfortunately the microgranites of Penmaenmawr have been severely quarried for roadstone, leaving a coastal mountain-side scored by workings and spoil heaps, equalled in their ugliness only by those of the Yr Eifl quarries on the Lleyn peninsula.

From the precipitous northern slope of this ridge there is a delightful view of Deganwy and Great Orme's Head across the sands and marshes of the estuary of the Conway river (Fig. 34). The steep and irregularly broken hills above Deganwy (on one of which the ruins of a Welsh castle stand) coincide with outcrops of volcanic rocks similar to those of Conway Mountain, but the Great Orme is made up of Mountain Limestone which in its light grey cliffs and crags makes a striking contrast with the scenery of the older rocks. Llandudno is built on a low spit of land (known as a tombolo) which has joined the former Great Orme island to the mainland. The long and narrow valley of the Conway, which is tidal to Llanrwst, has been carved out along the junction between the harder Ordovician volcanic rocks of Snowdonia and the more easily weathered Silurian shales and mudstones, which form the Denbighshire hills to the east (Fig. 34). The former course of the river, before it followed its present channel at Conway, can be traced from Glan Conway to Penrhyn Bay. Its diversion was probably caused by thick ice sheets which had travelled down the floor of the Irish Sea basin before impinging on the coast of North Wales. The marine shells dredged up by this ice sheet can frequently be seen in the boulder-clay sea-cliffs of the Denbighshire coast.

To the east of the Conway valley rise the Denbigh Moors, another great expanse of little-known mountain country built up by rocks newer than the volcanic group, roughly corresponding to the Upper Slates of the Lake District: most important among these rocks is the Denbigh Grit, which not only determines much of the character of the higher moors, but has also been utilized successfully as a building stone, especially at Conway Castle. Seen from a distance, the Denbighshire hills appear smooth-topped and monotonous, differing greatly from the more irregular craggy country of the volcanic tract. But the plateau of the Denbigh Moors is deeply dissected by many streams and in detail much of the area is extremely attractive, though it is much less known than most parts of North Wales.

Away to the east the Clwydian Hills really form an extension of the Denbigh Moors, for they have many similar features. But separating these two areas is another long tract of fertile lowland, much of it underlain by New Red Sandstone which imparts its own characteristics to this strip of country. Along it flows the river Clwyd, and several old towns, Ruthin, Denbigh, St Asaph and

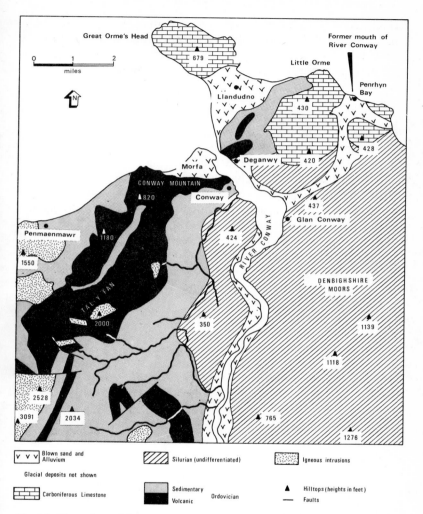

Fig. 34. *The geology of the Conway region. Note the contrast in elevation between the Denbighshire Moors (Silurian) and Snowdonia (Ordovician). Llandudno is sited on a tombolo.*

Rhuddlan, are spaced throughout its length. This green valley is a piece of the Midland plain which literally has been dropped amidst the Welsh uplands: it is bounded by faults, and is a miniature rift valley. The eastern boundary is especially sharp, a true fault scarp.

Anglesey

This brings us, however, into the borders of Wales, and it is necessary to turn back into the west, to make some reference to Anglesey,
the lowest county in North Wales. For Anglesey is a low plateau,
and most of its surface reaches to less than 300 feet above sea-level.
Broad flat-topped ridges rise to this height, and separating them
are shallow valleys cut by streams which flow in directions parallel
to the line of the Menai Strait. Here and there small areas rise
above the general level of good farming land but these isolated
hills are bare and rugged, the highest, Holyhead Mountain, reaching over 700 feet, but the remainder are nearer 550 feet. They stand
out as residuals above the lower plateau, but not because of any
special hardness, for these projecting hills vary as much in structure
as does the rest of Anglesey. They are relics of higher levels left
standing by denudation.

The almost uniformly low level of the Anglesey surface is the
more remarkable when it is remembered that the island is made up
very largely of extremely hard, old rocks (some of the oldest yet
encountered in this survey of England and Wales); great areas of
Anglesey are made of Pre-Cambrian rocks, often much contorted
by pressure and altered by heat; these metamorphic rocks include
gneisses and schists which may be paralleled in the Scottish Highlands. Besides these, granites, slates and limestones are also present,
the whole showing great complexity of structure. It is apparent that
the relief has little relation to the distribution of the rocks or to the
structure, save that the rivers have tended to etch out the weaker
belts, which trend, as do all the more important features in North
Wales, from north-east to south-west, the direction of the folds in
these highly disturbed rocks. The wide, level tract of Anglesey is
thought by some to represent a platform cut by the waves during a
period when the sea advanced over this whole area by cliff attack,

cutting across hard and soft rocks alike and truncating their structures. Looking across the Menai Strait from the island it is at once
apparent that the plateau levels of a few hundred feet above the sea
persist into the coastal parts of Caernarvonshire, for in a strip several
miles wide the mainland is more closely comparable with the island
than with the rest of the county (Plate 25). Above this level coastal
area, behind Bangor and Caernarvon, the mountains of Snowdonia
rise suddenly to heights of over 3,000 feet. An alternative hypothesis is to regard the 'Menaian Platform', as Dr Greenly called it,
as a surface from which a former cover of New Red Sandstone has
been stripped by erosion; this is known as a fossil land-surface,
similar to that of the Lewisian gneiss where the Torridonian sandstone of north-west Scotland has been removed.

Over the Anglesey surface the great variety of rock types (from
Pre-Cambrian to Carboniferous) is partly obscured by glacial
deposits, but its smoothing clearly preceded the Ice Age. Across
much of the island, ice moved from north-east to south-west in its
passage down the Irish Sea, though the line of the Menai Strait
was crossed by ice pushing outwards from the North Wales mountains. The main direction of ice movement shows itself, however,
in the arrangement of the drumlins, of which great numbers are
present in the north-west. Most of the area is under cultivation, but
there is much moorland, and in some parts the bare, rugged hills,
stone walls and houses, and dwarfed, wind-bent trees recall the
features of bleak uplands rather than of a low-lying, coastal region.
There are many small lakes, some true rock basins, like the Bodafon
tarns, others held up by glacial drift, and one, Parciau, formed by
the solution of an area of Mountain Limestone. Holyhead on
the bleak Holy Island, and now the largest town, is on the old
rocks of the extreme west, but Beaumaris, with its fine Edwardian
castle, has a more sheltered situation by the strait.

The Menai Strait calls for some attention. A channel some fifteen
miles long, it is bordered by low cliffs and by woods for much of
its length, and looks much like an enlarged river valley. It is probable that it represents a valley cut by glacial meltwater, similar to
those which run parallel to it on the mainland, but drowned in the
post-glacial submergence, as were so many Cornish valleys.

This short description can give only a hint of Anglesey's attractive coastline, the sandy bays and warrens, some backed by wide marshes, and the varied colours of the cliffs. The scenery of the mainland must be traced farther south into the Lleyn peninsula which has many affinities with the landscape of Anglesey.

Lleyn Peninsula and Tremadoc Bay

The conical hills which rise abruptly from the level of Lleyn are carved from volcanic rocks or igneous intrusions, similar in age to those of Snowdonia, of which this is a western limb. Two of the most striking are Yr Eifl, with its sheer coastal cliffs and intriguing Iron Age hill fortress of Tre'r Ceiri, and Carn Fadryn which dominates the western end of the peninsula. Here is a less-frequented part of North Wales, with hidden sea coves, fine cliffs and beautifully coloured beach pebbles (from the jaspers and serpentines amongst the Pre-Cambrian rocks). One is reminded once more of the treeless Anglesey landscape, although the western coastline has many resemblances to that of Cornwall, while at the eastern end of the peninsula is some of the most spectacular scenery in Britain, where ice-deepened valleys of Snowdonia have been invaded by the post-glacial rise of sea-level.

This is in the area overlooking Tremadoc Bay. The coast of this and neighbouring areas is of great interest, combining features of solid rock, glacial deposits, and sand and salt marsh, with alterations made by man (Fig. 35). The bay is an area of accumulation of beach material, mainly due to the prevailing south-westerly winds and the action of the waves caused by these winds sweeping material into the bay. These same waves also cause any beach of unconsolidated material to face as nearly as possible to their direction of approach. The beaches of the Lleyn peninsula are fixed between unyielding rocky headlands, but the crescentic boulder-clay cliffs of Porth Neigwl and the area between Pwllheli and Pen-ychan show this tendency well. Near the head of Tremadoc Bay are sandy beaches, notably Morfa Bychan on the north and Morfa Harlech (Welsh: *morfa* = marsh) on the south. The outer beach of Morfa Harlech faces almost square to the open sea, and dramatic

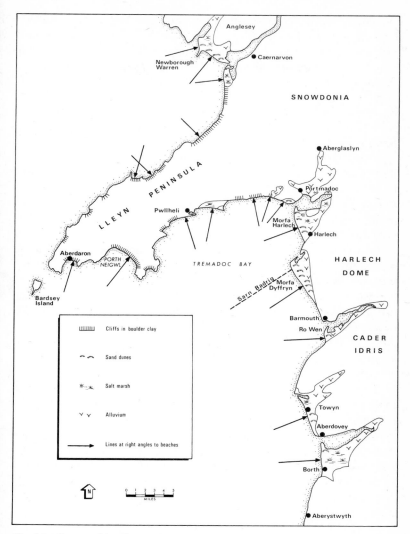

Fig. 35. *The coastal landforms of North and Central Wales.*

evidence for its recent growth can be seen by visiting Harlech Castle, built by Edward I in 1286. Like his other castles around Snowdonia, Harlech was formerly supplied by sea, and a harbour must have stood at the Water Gate. It is probable that in the thirteenth century there was a short spit here to provide protection for the harbour, but since that time the spit has lengthened, the dunes and marsh have grown, and now it is difficult to conceive ships moored below the castle on its solid rock cliff, with the famous view of Snowdon to the north.

Morfa Dyffryn, farther to the south, is similar to Morfa Harlech, with perhaps even finer dunes. At the northern end of Morfa Dyffryn is the former island of Mochras, made of morainic deposits. Passing seaward for about twenty miles from here is Sarn Badrig, or St Patrick's Causeway, several miles of which are exposed by a very low tide. This is one of a series of sarns, generally thought to be successive moraines of the ice sheets emerging from Snowdonia during the glaciation, but which are connected so strongly with the folk legend of the lost land of Cantref-y-Gwaelod.

South of Morfa Dyffryn again we have the Mawddach and Dyfi estuaries both guarded by spits, in these cases with their outer beaches of shingle, running nearly due north from the southern banks of the estuaries. Between these two estuaries are the marshes around Towyn.

Man's intervention in this scene is perhaps best seen at the head of Tremadoc Bay, where the embankment built by W. A. Madocks in 1811 enclosed the Glaslyn estuary and gained some rather poor farming land where the sea had formerly flowed almost to Aberglaslyn bridge. Present-day visitors may care to ponder on the scenic attraction of the area had the embankment never been built. Madocks also later built the slate-exporting harbour at Portmadoc, to serve the large Blaenau Ffestiniog quarries in the mountainland. The marshes of Morfa Harlech were also enclosed by an embankment at about the same time, in 1808.

The Country around Cader Idris

South of the rugged mountains of the Rhinogs, we pass into a region whose scenery was beloved by British landscape painters after its 'discovery' in the late eighteenth century. The dominant mountain here is Cader Idris, which rises splendidly from the Barmouth estuary in a great line of foothills surmounted by almost unclimbable precipices which face northwards towards Snowdon. Looking from the crest of this cliff, the irregular country which stretches down to the estuary at Barmouth, wooded in its lower part, presents quite different features from that which stretches away to the north, rising into the Harlech dome some eight miles away. To the right, Rhobell Fawr introduces yet another area of volcanic scenery.

Cader Idris itself is a mountain of volcanic rocks, comparable with those of Snowdon, the massive beds of extrusive lava being here augmented by great thicknesses of igneous intrusive rocks which were similarly forced up in a molten state but solidified as sills beneath the surface. Cader stands out because of the strength of these igneous rocks, which dip steeply to the south; it is really a great escarpment, trending with the run of the beds more or less east and west, and with its scarp facing to the north.

Once the volcanic group extended continuously northwards over the Rhinog Mountains to Snowdon, joining the great circle of craggy scarps which may be traced thence to Blaenau Ffestiniog and Dolgellau (Fig. 33). Arched up into a dome, the central area has been worn away, exposing in the centre between Harlech and Barmouth a thick series of grits and slates of Cambrian age. In this domed area the grits occupy the high ground, the massive beds giving rise to great steps on the desolate and almost uninhabited uplands; the slates produce softer scenery in a greener and more fertile belt which forms an almost complete ring around the Rhinog Mountains. Here denudation of the arched-up rocks has produced surface features in some ways comparable with those of the Weald, for although the rocks are much older and often harder, the hills much higher, and the whole area more barren, the general structure of an eroded dome is quite similar. Standing at the side of Traws-fynydd Lake, near the centre of the Harlech dome, a large nuclear-power station is regarded by many as an intolerable intrusion in the landscape. Its design and colouring, however, together with its relatively small scale against the towering hills, make one pause to wonder if this location is not eminently more satisfactory on aesthetic grounds than a similar power station sited on a low-lying coast such as Dungeness. After all, the roadstone quarries of Pen-maenmawr and the slate quarries of Llanberis and Bethesda are probably greater eyesores in the North Wales landscape. It is a sobering thought that to produce one ton of slate, up to twenty tons of waste are dumped on the Welsh mountain-sides.

Returning to Cader Idris, the view from the crest embraces the great ice-worn cwms beneath the summit. The steepness of the precipices which form the northern face of the mountain is due to the cliffs which back the cwms, whose basins, aided by moraine dams across their mouths, hold such lakes as Llyn y Gader and Llyn Cau. These great basin-like hollows on Cader Idris have often been likened to craters, especially by those who, knowing that the mountain is largely formed of volcanic material, have expected to find in it some shape resembling a volcano. Perhaps it is unnecessary to repeat here that Cader Idris is now a mountain simply on account of the hardness of its materials, and that its present form has no relation to any volcano from which its materials were originally derived. It is in fact an escarpment, with a scarp face to the north, differing only from the scarps of the Cotswolds and Chilterns in having igneous rocks as its more resistant beds. The main features of its relief were carved out long after the formation of its lavas and intrusions, and the basins on its scarp, the result of the Ice Age, are almost a thing of yesterday.

On the southern side of the mountain the volcanic rocks are succeeded by softer slates, mudstones and sandstones, which form the steep dip slope down into the Talyllyn valley. Southwards almost as far as can be seen from Cader extends the great plateau made of these rocks, belonging to the upper part of the Ordovician and to the Silurian, a rather monotonous expanse of even-topped hills when contrasted with the irregular and rugged summits of the volcanic tracts.

The main features of the Talyllyn valley will readily be recognized as due to glaciation, for it is a U-shaped valley holding a lake with alluvial flats at the head. The course of this valley has been largely determined, however, by the existence of a belt of shattered rocks coinciding with the line of a great fault, whose movements had led to the crushing of the rocks on either side. It differs from such faults as those at Settle and Cross Fell, in that the movement along it was horizontal rather than vertical; in the fault described in Giggleswick Scar (Fig. 23) one side was moved down relatively to the other, but in the Talyllyn fault the country on the south side has moved almost horizontally towards the east for a distance of about two miles. It thus represents a great wrench or tearing of the rocks, and the faulting is accompanied by considerable shattering along its course. Along this belt, therefore, stream erosion proceeded with unusual rapidity and the rivers which developed in it cut down their beds so rapidly that they captured the earlier drainage. To the east the same shatter belt is responsible for the trough in which Bala Lake is situated.

Accordingly the tributary streams in these areas have steep profiles for the main valleys have been deepened rapidly, partly owing to the uplift of the land and partly to the erosive work done by the ice; many tributaries thus hang above the larger valleys, and just before they join the main streams they often occupy narrow and deep gorges of great beauty cut into their older and wider high-level valleys. The Clywedog, as seen from the Torrent Walk near Dolgellau, is an admirable example of these rejuvenated streams, with a wooded gorge incised in the flat floor of the old valley, which provides a strip of cultivated land above the tumbling stream and beneath the barren crags. Higher upstream, the gorge diminishes and the wide valley is of simpler type, for the rejuvenation has not yet affected areas far from the over-deepened main valleys. Such features, however, have already been discussed in relation to the comparable examples in the Lake District, where many of the waterfalls and the most fascinating gorges owe their origin to similar causes. North Wales thus presents similar contrasts of bare rocky uplands, often dull in colour, and picturesque, wooded gorges, with rare expanses of flatter and more fertile lowland.

17. Central Wales

Seen from the crest of Cader Idris, Central Wales appears as a monotonous plateau rising to nearly 2,000 feet above sea-level, stretching through Cardiganshire to north Pembrokeshire and to Radnorshire. Lacking craggy mountains like those of the volcanic tracts, its hills are generally smooth in aspect, reaching so nearly to one level that from any high viewpoint the upland surface presents a surprisingly even aspect (Plate 27). There are few crags and in many wide areas on the plateau there are few places showing bare rock, partly because ice sheets were not present in Cardiganshire during the later stages of the Ice Age; over vast stretches the uniformity of relief and colouring is remarkable. Few habitations are found on these higher regions, and in many parts walls are only found occasionally, apart from the wall which usually limits the upland pastures.

Plynlimon is the highest mountain in this tract, and from its summit most of the essential features of the landscape can be seen. In fact, it is probably the best viewpoint in Wales. The ascent, for it can hardly be called a climb, is most easily made from the Aberystwyth–Rhayader road, where it crosses the watershed between the tributaries of the Rheidol and the Wye. Here a grey upland farm is situated in a wide, open valley which is almost devoid of trees, its sides covered by vegetation of that dull olive-green which characterizes so much of this country. There the little torrents have cut shallow, rocky gorges and bright clumps of gorse and heather add richness to the colouring, but for much of the way to the summit the path leads over poor pasture with patches of bog. As the track rises and the view expands, there comes a point when, a few hundred feet from the summit, the whole expanse of the

Plate 27. *The plateau of north Cardiganshire, Central Wales. The view shows the country south of Plynlimon, where the High Plateau (1,800–2,000 feet) is well preserved despite the incision of the headstreams of the river Wye.*

plateau becomes visible for miles to the south and east; the flat-topped hills, reaching a common level at which we are standing, seem to form a vast plain out of which Plynlimon slightly raises its head.

But while Plynlimon may thus be one of the easiest mountains in Wales to ascend, its northern face has none of the gentleness of the other slopes, for beneath the summit a little lake lies in the shelter of a deep cwm, from which rise steep, grey cliffs. Here, as is so frequently the case, ice erosion has led to the production of precipices in otherwise subdued country, and these effects are most pronounced on the north or north-east facing slopes.

On a clear day the view from the summit of Plynlimon embraces parts of nearly every county in Wales, and of the Welsh borders. The level plateau stretches northwards to the shoulders of Cader Idris and Aran Mawddwy; these mountains and the higher peaks of Snowdonia rise from it like rugged hills upon a plain. Turning to the south, the plateau forms a remarkably accordant summit surface, the valleys which dissect it being hidden from view. Just visible to the south is the crest of the Brecon Beacons, which is sufficiently raised above the plateau to prevent any glimpse of Glamorgan. Near at hand, the level falls quite rapidly to the west, and along the shores of Cardigan Bay, visible from end to end, extends a plateau at a lower level, traversed by the narrow valleys of rivers which flow into the bay, and much of it covered with a multi-coloured patchwork of fields.

There are in fact three main plateau levels in this region, which Professor E. H. Brown has named respectively the High Plateau, the Middle Peneplain and the Low Peneplain (Fig. 36). The main features of the High Plateau have already been described, but owing to their lower elevations the Middle Peneplain (1,200–1,500 feet) and the Low Peneplain (800–1,000 feet) possess greater expanses of improved land and scattered farms. It may be useful to indicate what is known about the origin of each before going further with the description of the region. In the first place, however, it must be emphasized that the rocks which build these plateaux are identical in all respects; they consist of slates, mudstones and sandstones of many types but frequently of a grey colour. All of these rocks have been subjected to folding and faulting, and the whole area is of great complexity. It is clear, therefore, that the plateau surfaces do not reflect the geological structure, but have been cut across harder and softer beds alike. They are therefore surfaces produced by denudation, and it is in the history of their erosion that their origins must be sought.

The High Plateau was regarded by Ramsay, who long since

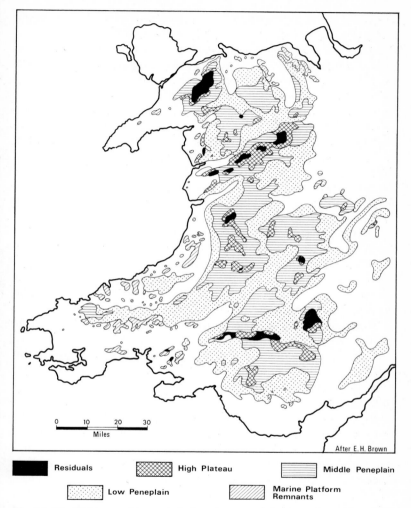

■	Residuals	▨	High Plateau	▤	Middle Peneplain
░	Low Peneplain	▨	Marine Platform Remnants		

0   10   20   30
Miles

After E. H. Brown

Fig. 36. *The Welsh peneplains and platform remnants (after E. H. Brown).*

described its chief characters, as the result of wave action at a time when the area stood some 2,000 feet lower in relation to the sea. Although most writers have accepted the view that this land was planed down when it stood near sea-level, attributing its uplift to a later date, there has lately been a tendency to regard the levelling rather as the work of meandering rivers and their tributaries. Generally speaking, the older geologists were more ready to attribute features to marine erosion, but later work, particularly that associated with the late W. M. Davis, has led to the recognition of the adequacy of rivers and weathering agents to produce peneplains if sea-level remains nearly constant for sufficiently long periods. There is thus room for some differences of opinion regarding the origin of the High Plateau, although Professor Brown believes that rivers almost certainly played the major part in its planation. He goes on to trace the gradual evolution of the Welsh drainage pattern and shows how river encroachment and periodic uplift of the land have led to the formation of the two lower peneplains. Each of these has grown in size only by the subsequent destruction of the older and higher plateaux, the remnants of which are left as residuals standing above these extensive erosion surfaces.

Some geologists have held that the chalk sea extended over the greater part of Wales, and have seen in its advance the principal cause of the levelling of the High Plateau. They couple with this belief the view that Chalk was also deposited (unconformably) over the whole area, and that the south-easterly direction of many of the rivers resulted from their initiation as consequent streams on a surface of Chalk raised from beneath the sea and gently inclined in that direction. The rivers, having removed the chalk cover, have become superimposed on the complex structures beneath, in a manner similar to that discussed in regard to the Lake District rivers (although there was no striking radial arrangement in Wales). The superimposed character of the drainage is widely accepted, but the former occurrence of the Chalk is disputed, and it may be that the peneplain of the High Plateau is much more recent than the Chalk, as Professor Brown believes.

At an elevation of about 600–700 feet a marked platform can be seen flanking the coast of Wales, especially in Cardiganshire. There is little doubt that it was formed at a later date than the higher plateaux although its origin is again open to dispute. Professor O. T. Jones believed that it was planed down by a west-flowing system of streams which reduced it to a height not far above the sea-level then existing. Professor Brown, on the other hand, is convinced that here at last is evidence of an old marine shore-line, slightly older than the first ice sheets in the region. Subsequent uplift has led to the rejuvenation of the rivers and they have thus incised deeper valleys in the surface of the coastal plateau. In this tract also there are thus large expanses of fairly level upland, but whereas many valleys of the upper plateaux are wide and open, the coastal plateau is crossed by narrow valleys which are often picturesquely wooded; where the valleys have wider floors, as in the case of the Rheidol below Devil's Bridge, they are occupied by cultivated land, but cultivation is generally confined to the platform surface. Here the farms have both arable and pasture land, and there is a surprising amount of corn grown, especially oats in many places; above 600 feet, on the other hand, the proportion of pasture is greater, for the growing season is much shorter. It is only in some of the highest farms of the upper plateaux, however, that the pasturing of sheep is the sole occupation. Formerly these farms were used mainly for summer pasturage of stock from the lower levels, and there is still a considerable transference of sheep for summer grazing.

Within the area of the higher plateaux there are few villages and such as occur are placed in the valleys; above these there are only scattered farms, and great tracts are almost uninhabited. Between the valley of the Teifi, with its little market towns, Tregaron, Lampeter and Llandyssul, and that of the Towy, with Llandovery and Llandeilo, is one of the biggest tracts of bleak upland in southern Britain. About twelve miles across and twenty in length, it is almost everywhere above 1,000 feet high, merging northwards into the Plynlimon plateau. Much of its drainage goes southwards into the Towy or to the Wye system. In the higher parts the valleys are wide with gentle slopes passing gradually into the plateau surface, but as the streams approach the borders of the upland, their valleys steepen into gorges. In the case of the Towy itself this change occurs near Rhandirmwyn. The uplift which caused the rejuvenation of these streams has only enabled them so far to deepen their valleys in the lower reaches; it may be expected that in time the gorges will be cut back into the higher upland, but the uplift is too recent in date for this to have occurred so far above the sea.

Within the uplands, apart from the few rough tracks climbing sharply up from the valleys, there are no roads and scarcely any habitations. There are in fact few more desolate areas in Wales. To many, these higher uplands may seem monotonous, and they are not likely to attract any except those who appreciate clean air and open moorland.

The towns of this region, so closely related to the main valleys, are not usually so attractive as the small market towns of England; when seen from a distance, however, they often fit very appropriately into their background, for the use of local stone ensures a suitable if dark colouring. The stone is usually in large blocks, with little ornament, but the mortar is sometimes picked out in white with rather a startling effect. Many cottages, however, are regularly washed over with white or a very attractive pink, and grouped picturesquely on the hillsides they are then bright and conspicuous. Few cottages are thatched, most have heavy slates, but it is regrettable to find that even so near the home of slates the use of galvanized roofing and manufactured tiles is spreading into many villages.

Only rarely do churches add to the beauty of the villages, for the parishes are large and the churches few. More frequent are the bare, rectangular chapels, so characteristic of most of Wales, and often less beautiful than many a Cotswold barn.

Stone walls are found in some areas, but in many places, especially on the coastal plateau, the lanes are lined by rough walls of vertically piled slates, above which rise high, grassy banks, often bright with flowers: in late summer great masses of yellow ragwort dominate the colour of the landscape. Generally, however, the higher plateau lacks these striking colourings, for its poorer drainage qualities lead to the development of extensive tracts of peat moor with drab brown or olive colouring.

Where the higher plateaux meet the coastal plateau the rivers are often incised in deep gorges of great charm. None is more famous, or more deserving of fame, than the glorious valley at Devil's Bridge. Near here the Rheidol, leaving the wide valley which it follows from its source in the cwm-lake under Plynlimon, plunges into a narrow rocky gorge and takes a sharp turn to the west. Here in fact are two valleys, belonging to two different rivers, for the upper part of the Rheidol was formerly a south-flowing stream of the High Plateau, which has been captured by the rapid headward erosion of a powerful stream, with a steep gradient occupying a deep valley on the coastal plateau. The fine chasm, with 'woods climbing above woods', seen from the front of the hotel at Devil's Bridge, has been cut down several hundred feet lower than the wide, old valley on the floor of which runs the road to Ponterwyd. Into this deep gorge the little river Mynach also crashes in the magnificent series of falls at Devil's Bridge, the narrow cleft through which this river flows and the pot-holed bed being typical of the work of an energetic stream.

The Teifi valley also has many attractive features. Near the great bog of Tregaron it has the width and gentle slopes of a mature valley, but it narrows before reaching Llandyssul, and thence to the sea its course is through a deep and picturesque valley, often with steep, wooded slopes by Henllan and Newcastle Emlyn. At Cenarth, above the narrow grey bridge, is a small but attractive fall and the river is actively deepening its valley with an energy which is surprising when it is remembered that this place is little more than ten miles from the sea. Still lower, where Cilgerran Castle stands on its rocky cliff, the wooded gorge is even more striking, while at Cardigan, where the river is tidal, and where the castle stands above another old bridge, the valley is deeply incised in the coastal plateau, and the village of St Dogmael's scrambles up the steep slopes.

The smaller valleys crossing the coastal plateau are just as pictur-esque. Many of them are so narrow that they hold no road save where old villages are situated on the coast at their mouths or a mile or more inland. The bright, colour-washed houses of Moyl-grove, hanging on a steep valley-side west of Cardigan, and Aber-porth and Llangranog on their little bays, are typical of the gems of this coast. The main road between Cardigan and Aberayron keeps some way from the coast along the more level plateau top, avoiding the sharp gradients of the narrow valleys in which the roads are often unpleasantly steep. Most of the coast can only be seen on foot; it has few great cliffs, and it appears as if a smoothly moulded, green country has been chipped by the waves along its edge, revealing the rocks which built this region in all their variety of colour and form. At Aberporth are cliffs of silver-grey slate, while at Llangranog dark and highly disturbed shales stand out in stacks of grotesque form.

Leaving the coast for the eastern part of Central Wales, we may turn to the green and fertile vale of the Towy, with its flat, alluvial meadows through which the river meanders from Llandeilo to Carmarthen. Here is a tract quite unlike that of any other part of the region described, a wide plain extending far inland, whose accessibility and richness are marked by a line of finely placed castles.

This same structural line extends still farther to the north-east, through Llanwrtyd and Builth to Llandrindod, where anticlinal folding has brought to the surface volcanic rocks belonging to the same Ordovician group as those which built Cader Idris and Snowdon. The reappearance of these rocks, deeply buried under the Central Wales plateau, introduces into this tract something of the variety and ruggedness of hill outline found in North Wales, features which add not a little to the attractiveness of the 'Wells' country.

Finally a mention should perhaps be made of the reservoirs of this country. This is no place to enter into political and social controversy, but it should be pointed out that the ample surplus of rainfall over evaporation, the impermeable nature of most of the rocks and the occasional presence of narrow defiles for dam sites with wide glaciated valleys above for the reservoirs themselves, make ideal locations for water catchment to supply the distant cities of England, albeit at the expense of upheaval to the sparse rural population. Fashions in reservoir construction as in everything else are changing, and modern reservoirs, such as Llyn Celyn on a tributary of the Dee, and the Clywedog reservoir on a tributary of the Severn, are used to control the flow of the rivers, thus alleviating distress farther downstream caused by floods and low flows, as well as making the reservoirs accessible for recreation. This is in contrast to the older Elan reservoirs supplying Birmingham, and Lake Vyrnwy supplying Liverpool, which have aqueducts direct to the cities from the reservoirs.

18. South Wales

Traditionally South Wales was entered along the narrow coastal plain of Monmouthshire, following the Roman road which links Gloucester (Glevum) with Caerwent (Venta Silurum) and Caerleon (Isca). Today, however, many visitors arrive by way of the Severn Bridge motorway which affords spectacular views of the river Wye, the Forest of Dean, the higher Welsh hills and the Severn estuary which can be traced away to the west towards the cities of Newport and Cardiff. The latter stands where three rivers from the hills enter the estuary, the Ely, the Taff and the Rhymney, two of which meander amazingly through the low estuarine flats of soft grey mud. The earliest settlement, on a river terrace above these flood plains, marked the crossing of the Taff by the most important route into South Wales, on a site used successively by Romans and Normans. But the city now extends in several directions on to higher ground. Eastwards it spreads up the slopes of Penylan, where a small patch of Silurian limestones and mudstones stands out above the surrounding plain.

From this viewpoint the scenic features of the region can be picked out. Away to the north rise a series of hills marking the borders of the coalfield, backed by bare uplands of over 1,000 feet in height; in that direction the mountain scenery is typically Welsh. Nearer to Cardiff, however, and extending westwards almost to the shores of Swansea Bay, is an area which scenically more nearly resembles the English Midlands, an undulating tract of greater fertility recognized long ago as one of the richest agricultural regions in Wales.

The Vale of Glamorgan

This fertile coastal tract is called the Vale of Glamorgan, but it is not a valley in any sense, for several small rivers drain its different parts. Varying in height up to just over 400 feet, it has wide areas at 200 feet or thereabouts and may be regarded as a dissected plateau, its surface having been cut, possibly by the sea, from a variety of Mesozoic rocks which in general have resisted denudation much less than the older rocks which make up the South Wales coalfield to the north. Whatever its origin, this coastal tract ends abruptly at the shore-line in cliffs which are nearly always vertical. The land is obviously being attacked continuously and with success, for weathering scarcely manages to reduce the sharpness of the cliff-top before another fall produces a new steep face.

The cliffs along this portion of the coast, from Penarth to Porthcawl, are never much above 100 feet high, but they can rarely be climbed, and for miles at a time they present a continuous front to the sea, broken only where cliff recession has cut back into coastal valleys. In front of the cliffs is an extensive platform cut in the rocks by the waves, in places heaped up with debris of boulders from the destruction of the cliffs, elsewhere swept bare by every tide. It is from this tract beneath the cliffs that the beauty and interest of this coast can best be appreciated.

Penarth Head shows at once the rocks which determine the character of the area; beds nearly horizontal succeed one another regularly up the face of the cliff in striking bands of colour, red marls with tinges of green, pale green marls, black shales and blue and yellow limestones. The sequence is exactly that which underlies the Vale of Gloucester, red marls of the New Red Sandstone followed by the limestones and clays of the blue Lias. Southwards from Penarth to Lavernock the red and green marls form much of the cliffs and foreshore, but gentle undulations of the beds form a wide syncline and bring blue limestones down to sea-level, the harder beds making small but distinct headlands at either end. Although red rocks once more form the cliffs beyond St Mary's

Well Bay, the greater part of the coast of the vale is occupied by Lias limestones, the regular, thin beds being cut by vertical joints and often gently inclined so that stronger and weaker groups alternately form cliffs of different resistance.

In its cliff features this attractive coastline is strongly reminiscent of Whitby, where similar rocks occur, but it differs greatly in the siting of the villages. There are few villages actually near the sea, and most are quite invisible from the shore. The grouping of cottages along the sides of every bay, so typical of the Yorkshire coast, is lacking in Glamorgan, for although this area is quite heavily populated when compared with much of rural Wales, the inhabitants of the villages depend on agriculture more than on the sea. Thus the villages appear to turn their backs to the sea and are often found below the skyline in the valleys a short way inland, and are spaced at intervals of only a mile or so. The remarkable early history of Llantwit Major, the 'University' of ancient Wales, illustrates the degree of settlement before Norman times: castles too are numerous, for the richness of the land was appreciated by the Normans, and its conquest was completed before the end of the eleventh century. Many of the villages are pretty, the cottages often built of the grey-blue Lias limestone, sometimes colourwashed, and dominated by small, grey churches with tall, square towers.

This unspoiled green 'Vale' is the more remarkable since it is in such close contact with the densely populated industrial regions along its northern borders. For apart from the growth of Bridgend, where the coalfield valley of the Ogmore comes out to the coastal plateau, and of Porthcawl, a modern resort by the sea, the greater part of the vale has remained essentially unaffected by the changes of the Industrial Revolution. Cowbridge, for long its principal town, avoided the influence of the main railway line (routed some miles to the north) and remained a quiet country town until the return of road transport led to rapid and unpleasant changes in its main street, only partly alleviated by a modern by-pass. But away from the great road to the west, most of the vale is still quiet and undisturbed. The country around Barry is the most important exception: here a hamlet became a large town almost in a year, following the construction of docks when the coal export trade was developing most rapidly. Probably no town in Britain grew up more quickly, from a population of about 100 at the 1881 census to 13,000 in 1891.

The coast at Barry shows some striking differences when compared with that of most of the vale, for pointing to the south are three rocky headlands quite different in aspect from the red marl cliffs of Penarth or the regular Lias cliffs at Rhoose and the area to the west. Built of thick and steeply dipping beds of massive grey limestone, belonging to the group of the Mountain Limestone, these cliffs are capped in several places by the basal members of the New Red Sandstone, nearly horizontal red rocks thus resting unconformably on the terraced surface cut in the steeply inclined limestones. In short, at Barry the base of the rock series which floors most of the Vale of Glamorgan (that is the New Red Sandstone and Lias) is seen above sea-level, exposing the older and more disturbed rocks beneath. Similar conditions recur near Southerndown; from Nash Point to Dunraven the cliffs consist of regular Lias beds except that Witches Point is a rugged headland formed mainly of Mountain Limestone, which reappears along the foreshore beneath Sutton and in the low cliffs at Porthcawl. Inland the irregular surface of the older rocks is still higher in places and the newer beds have been removed so as to expose tracts of Mountain Limestone which determine the character of small but characteristic limestone regions in the vale, e.g. between St Bride's Major and Ogmore. Formerly the new rocks had a still wider extent, and some of them may have stretched far to the west and north, but from most areas they have been removed by denudation. Most of the rest of South Wales is thus built up of older rocks.

The Coalfield Borders

These areas may best be approached by going northwards from Cardiff, traversing first the wide tract of undulating country around Llanishen and Lisvane, which is made up by a group of red marls similar in aspect to those of the New Red Sandstone but belonging to a much older series, underlying the Carboniferous rocks and known as the Old Red Sandstone. Above Rhiwbina the ground rises to an escarpment which in turn is succeeded by a dip slope, and then by two other escarpments. These features are narrow, with steep dip slopes, a result of the high dip, and they trend approximately east and west, the structure being similar in many respects (except rock type and age of folding) to the Hog's Back of Surrey. They give rise to a pleasantly varied tract of country, wooded on some of the higher ground and on many steeper slopes, with narrow strips of more fertile country excavated along the less resistant beds. The nature of the ridges and their formation by an alternation of relatively hard and soft rocks is indicated on the accompanying diagram (Fig. 37).

The first ridge, which extends to the heights on which Ruperra Castle stands, is made up of tough conglomerates, with white quartz pebbles set in a red matrix, the uppermost part of the Old Red Sandstone. The second ridge, familiar in Cefn On and extending to the hill above Castell Coch and the beautiful Garth Wood or Little Garth, results from the Mountain Limestone, which gives rise to characteristic light-grey crags; the depression beyond cor-

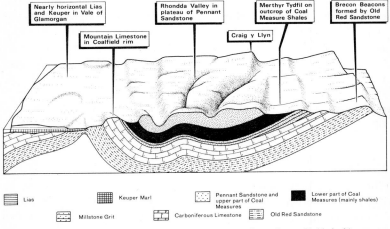

Fig. 37. *The structure of the eastern part of the South Wales coalfield, looking westwards.*

responds to the outcrop of shales, the lower part equivalent to the Millstone Grit of other areas, the upper part being the lower portion of the Coal Measures. North of this again is a scarp which rises above Taffs Well into Garth Hill and Mynydd Rudry, the dark barren summits of which, covered by coarse grass and bracken, form an impressive contrast to the wooded ridges farther south: these higher hills, situated just within the coalfield border, are very reminiscent of the grit moorlands of the Southern Pennines. They are built by a group of dull grey-green sandstones, the Pennant Sandstones, which crop out over a great part of the surface of the South Wales coalfield and are responsible for most of its more distinctive landscape characters (Fig. 37), in addition to their extensive use as a building stone.

These successive ridges of the coalfield border are breached by the rivers· which flow across them. The Taff, for example, has cut fine gorges at Taffs Well and Tongwynlais where it crosses the harder beds, but the valley changes in width as it passes from harder to softer beds. In cutting for itself this exit from the coalfield the river has lowered its bed, but has been unable to widen its valley where the rocks are hard, although where the rocks are less resistant the valley has been greatly enlarged, often with the aid of short tributary streams. So in parts of the valley there is scarcely room for the railways, canal and road, while elsewhere the alluvial plain spreads considerably and villages find more extensive sites.

To the west, these border ridges are generally less distinct, but eastwards around the coalfield rim to Pontypool they are very conspicuous. and where they are cut through by the rivers, the Rhymney at Machen and the Ebbw at Risca. the conditions already described in the case of the Taff gorge are reproduced.

The Coalfield

The main part of the coalfield, however, is quite different from these sharp border ridges. Wherever the Pennant Sandstone forms the surface the tract is high and barren, the smaller areas occupied by the shales which occur below (and in a few places above) the sandstone leading to rather more fertile and generally lower areas.

Thus Caerphilly is situated in a wide basin of coal-bearing shales, where the Rhymney escapes from the confinement of the hills and almost loses its way to the sea, while all around, the smooth bare dip slopes, marked here and there by great black spoil heaps, rise to the Pennant uplands. This fertile tract, though within the coalfield, is rather of the lowlands than of the uplands, and hence it was selected as a site for a great castle comparable with those near the lowland borders, for Norman influence penetrated but slowly into the upland plateau regions. This splendid ruin, built of the green-grey stone from the Pennant rocks, is almost in the straggling main street, with its old and rather picturesque stone-built houses. Caerphilly was one of the earliest towns to grow up within the coalfield, and it developed rapidly in the early days of the coal trade when the shallow seams were mined extensively. These were of limited extent and were rapidly exhausted, and as the more important seams of the coalfield here occur at great depths, industrial development has been almost arrested in this area.

Most of the coalfield area of Monmouthshire and east Glamorgan is occupied by a high plateau, rising from about 1,000 feet on the southern border to just under 2,000 feet in Craig y Llyn near the head of the Rhondda valleys (Fig. 37). This tract is one of the most populous areas in Britain owing to the great development of coal-mining, more particularly in the last century or so, but the high uplands show little evidence of such a change, for the towns are confined to the valleys and the open moorlands remain bleak and almost uninhabited. The scattered hamlets and shepherds' huts, connected by ancient trackways across the hills, date back to times before the Industrial Revolution, when the valleys were densely wooded and often marshy and impassable. Here and there on this almost level upland are old village churches, few in number because the old parishes in this sparsely peopled area were of great size; nowadays many of these churches stand on the lonely uplands right above the crowded industrial valleys. These hill-tops are often remarkably flat, for the valleys are deep, narrow trenches cut into the plateau surface, and from many points the skyline appears to be level and scarcely broken, the towns of the valleys being quite invisible. The plateau surface is probably continuous with the High

Plateau of Central Wales, and represents a peneplain formed at the same time, but as in the case of that region, the exact mode and date of origin are open to some doubt.

The valleys which are cut deeply across this plateau are more familiar than the upland itself, for the excessive ugliness of these industrial towns made them notorious in a generation which learned something of town planning, while their still more pitiable condition during the depression of later years has attracted even more sympathetic attention. These valleys which drain into the sea at Cardiff and Newport extend far into the hill region, sometimes commencing beyond the northern boundary of the coalfield; the Rhondda Fawr and Rhondda Fach, the Cynon, the Taff, the Rhymney, the Sirhowy and the Ebbw all follow nearly parallel courses from the north-north-west, flowing swiftly along their steep beds to the sea. The simplicity of this arrangement of the rivers is at once suggestive of a series of consequent streams. But the rivers are not following the dip of simple cover rocks, and if their courses were determined by such conditions, it is clear that the gently dipping beds have all been removed and the rivers superimposed on older rocks of more complex structure. Thus the rivers now cross various structures on their way to the sea, sometimes flowing with the dip, sometimes against it. The presence of Mesozoic beds covering these older structures in the Vale of Glamorgan is notable in this connection, and some postulate that Mesozoic rocks formerly stretched over the coalfield and formed the surface on which the simple river system was first developed.

These coalfield valleys often have steep sides, rising sharply from a narrow strip of more level ground; the main road winding along the valley is flanked by almost continuous works or shops, one town or village joining with the next, each having no centre, no beginning and no end. The dearth of flat ground has led to the problem of colliery spoil heaps being situated on the valley slopes above the dismal terraces of grey houses, often built of Pennant Sandstone (Plate 28). The steepness of the slopes and the instability of the tips have been responsible for much slumping, culminating in the horrifying Aberfan disaster in October 1966. The pressures of re-housing and the lack of suitable building land has led to an

Plate 28. *Hopkinstown, near Pontypridd, South Wales. Large parts of the valley floors in the coalfield are occupied by colliery buildings and railway tracks. Many of the domestic settlements, frequently built of the local Pennant Sandstone, have been sited on the steep valley sides. The lack of space on the valley floor has also led to the siting of the colliery spoil heaps high up on the interfluves between the valleys, often immediately above the terrace housing.*

entirely new departure in the siting of residential settlement. High up on the plateau between the two Rhondda valleys the new settlement of Penrhys is slowly taking shape at an elevation of 1,000 feet O.D.

The Rhondda valleys are bottle-necks starting among the highest hills of the coalfield, and for long there was no route out to the north, but lately magnificent roads have been made from one valley to another, and those linking the Rhondda with Hirwaun and Cymmer afford wonderful views of the plateau. The more easterly valleys, however, open out before they reach the plateau, for the shales of the lower part of the Coal Measures have there been worn more quickly than the Pennant Sandstone which hems in the valleys just below. Here the readily accessible coal and ironstones led to early developments, and the bigger and rather older towns which form a chain along this belt, Brynmawr, Merthyr Tydfil and Aberdare, have grown on more spacious sites than those of the narrower valleys (Fig. 37), but the more quickly exhausted mineral wealth has left some of them even more desolate.

South of this subdued belt the Pennant Sandstone rises in a steep scarp, most prominent in the west, where Craig y Llyn towers in dark and nearly vertical cliffs above two cwm lakes. From the crest of this ridge the view to the north embraces a series of ridges corresponding with those which form the coalfield borders on the south; here, however, the dip of the rocks is to the south and is more gentle, the rocks forming wider outcrops and higher scarps. That of the conglomerates at the top of the Old Red Sandstone is the most impressive. From Craig y Llyn we see the gentle dip slopes of dull green pastureland, peat bogs and cotton-grass moors rising to the crest of Fforest Fawr and the Brecon Beacons; these, like Craig y Llyn, have mighty scarp slopes facing north, almost as high as Cader Idris and many of the more famous mountains of the north, exposing great cliffs of horizontally bedded red rocks. This line of hills stands out above the peneplain of the coalfield and Central Wales, much as do some of the mountains of North Wales. It affords a splendid view of the country around Brecon, an undulating region of rich greens and deep red fields comparable with that just north of Cardiff.

The arrangement of the various rocks of the coalfield borders is thus similar in both north and south, for the coalfield is a basin with the rocks older than Coal Measures outcropping in regular sequence around the margins (Fig. 37). Westwards these northern ridges continue beyond the heads of the Neath and Tawe valleys, the conglomerates of the Old Red Sandstone capping the precipitous scarps of Fan Gihirych and Fan Hir; the last-named rises in a sheer cliff above the cwm which holds the moraine-dammed lake of Llyn y Fan fawr. But west of this area the conglomerates become thinner and this ridge diminishes rapidly in stature, so that between Ammanford and Llandeilo it is relatively inconspicuous. As the conglomerate scarp diminishes, however, another ridge rises to the south of its line to form the great moorland of the Black Mountain, its bare and stony dip slope emerging from beneath the coal-bearing shales of the Amman valley to the crest of a scarp. This mountain is built up mostly by tough white quartzites which here make the Millstone Grit more like its equivalent formation in Yorkshire than is usual in South Wales.

The coalfield west of Neath is less elevated than farther east, partly because the Pennant Sandstones here contain greater proportions of shale. Thus while there are many bare and dreary upland tracts, the valleys on the whole are wider and are cut down more nearly to sea-level. Consequently their tributaries have been able more actively to erode courses on the softer shale outcrops, and the drainage shows a greater harmony with the geological structure than it does near Cardiff. For example, the Loughor for much of the way from Ammanford to the sea meanders in a wide and often marshy valley, enabling its tributary the Amman to cut a great valley along the strike of the shales of the Coal Measures. Around the Loughor estuary, too, wide valleys are excavated in the softer shales. So the harder sandstone bands stand out in prominent ridges, like that which fronts Swansea Bay. This is breached at several points, but notably by the Tawe, which occupies quite a narrow gorge between Town Hill and Kilvey Hill just before it enters the sea at Swansea.

This steep scarp, rising to about 600 feet within a short distance of the coast, has greatly affected the growth of Swansea, and although terraces now extend over most of Town Hill, the sister hill on the east is still almost bare, the town creeping round its flanks. On the north of the town metal-works have produced a dreary desert, and the view of Landore from the railway is one of the most depressing in Wales, except perhaps at sunset when the glowing river winds among the dark masses of derelict works. Almost in contact with these relics of industry are little hill farms and whitewashed cottages which have somehow contrived to exist through the period when fumes obliterated nearly all vegetation from the hillsides; with the decline of copper-smelting some grass has lately been encouraged to grow. But in its southern and western parts Swansea shows far less the presence of industry, and spreads, a pleasant and friendly town, around the sandy shores of a wide bay. On the eastern shores of Swansea Bay the fine coastal dunes which stretch northwards from Kenfig Burrows (with its buried medieval settlement) are now being rapidly destroyed by the encroachment of the vast Port Talbot steelworks, the pollution from which is rapidly killing the woodland on the hillslopes behind the town.

The Vale of Neath

The most important rivers entering this bay, the Tawe and the Neath (or Nedd), flow in valleys quite unlike those of east Glamorgan, for they are almost straight trenches, open for many miles inland, running from north-east to south-west. Both of these valleys have been determined by belts of crushed and weakened rocks, and are thus in some ways comparable with the Talyllyn-Bala valley of North Wales. Along these easily eroded belts the rivers have cut down rapidly to near sea-level; their tributaries, however, have had to contend with normal rocks, and they occupy high-level valleys from which they tumble into narrow gorges before joining the main valley. These effects have been increased as a result of glaciation, for much ice from the Brecon Beacons reached the sea along these valleys; during their retreat, the valley glaciers built up recessional moraines across the valleys, that at Glais, north of Swansea, being the most impressive.

In the Vale of Neath, industrial development is so localized that great stretches of really attractive country remain. At many points the river Neath flows in a narrow alluvial plain, but the valley sides are steep, especially where they are cut in the Pennant Sandstone. About half-way up the valley, near Resolven, several torrents fall swiftly into the main stream, making picturesque waterfalls in their rapid descent; that at Melin Court is the largest, the stream falling over a ledge of massive sandstone into a short wooded gorge left by the receding fall. In North Wales such falls would be visited by thousands, but in the south they are little cared for, and it is not uncommon for refuse to be deposited in the most beautiful of the gorges.

Still farther upstream, near the head of the Vale of Neath, the river scenery is even finer, for at Pont Nedd Fechan (or Pont Neath Vaughan) several streams occupy gorges of extraordinary beauty, and give rise to a series of waterfalls which have become deservedly famous. Farther upstream other attractive scenic features are shown where some streams cross the outcrop of the Mountain Limestone, as in the case of the Mellte south of Ystradfellte, which plunges into a cave (Porth yr Ogof) beneath a limestone cliff, leaving the old bed dry, until it reappears some distance below.

Gower

Away to the west of Swansea stretches the peninsula of Gower, a low coastal plateau for the most part outside the coalfield and thus, like the Vale of Glamorgan, bringing a pleasant rural area into the closest proximity to the industrial regions. But Gower is more isolated than the vale, through which passed the old routes to the west; Gower is so deeply cut by the Loughor estuary on the north-west that it forms a cul-de-sac, and it has escaped many of those influences which modified the vale. It has no railway and no towns, and is almost shut off from the Welsh-speaking areas in the adjacent coalfield by the scarp of the Pennant Sandstone which has long formed a barrier between the two regions.

The Gower peninsula consists of sharply folded Old Red Sandstone and Carboniferous rocks, forming anticlines and synclines

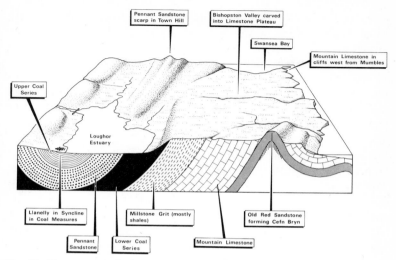

Fig. 38. *The structure of eastern Gower, looking east.*

running nearly east and west (Fig. 38). Small patches of New Red Sandstone here and there, and a red staining of the rocks in many places, indicate that the newer rocks found in the vale once spread over much of this region, but if this were so, they have almost completely disappeared. Structurally the Gower peninsula may thus be compared with the basement of the Vale of Glamorgan.

The relief of the peninsula shows some relation to its structure, for the anticlinal areas where Old Red Sandstone is brought to the surface all stand up as hills about 600 feet above sea-level, but apart from this the peninsula is a low plateau. Their resistance to denudation is mainly due to the presence of quartz conglomerates identical with those which form the summit of the Brecon Beacons. Running almost through the middle of Gower, the narrow ridge of Cefn Bryn marks the position of a sharp anticline whose steep slopes to north and south are littered by irregular blocks of the tough conglomerate. From the old track which follows the crest of this bracken-covered hill the best views of Gower may be obtained. Cefn Bryn and its westerly extensions in Llanmadoc Hill and

Rhossili Down then stand out as conspicuous residuals of nearly equal height, above a smooth plateau in which the rivers have carved narrow trenches.

In many of its features the plateau surface of Gower recalls those of Anglesey and the Vale of Glamorgan; from any good viewpoint the skyline is remarkably even, the surface cutting across the edges of the upturned rocks and reflecting little of the underlying structure (Fig. 38). Large tracts at just over 200 feet and just over 400 feet may be recognized in all these areas, and it is likely that they were cut by wave action in two stages when the land stood about 200 feet and 400 feet lower in relation to the sea. The elevation of these platforms appears to have been uniform around the coasts of Wales, and accounts for the characteristic smoothness of many coastal areas.

The greater part of southern Gower is made up of Mountain Limestone, which produces, at any rate in the valleys, many of the characteristic features of limestone scenery. The narrow valley which stretches from Bishopston to the sea at Pwll du is the best-known example, with a stream plunging underground as it reaches the limestone outcrop, to rise again near sea-level; a pleasant valley of white limestone crags and rough wooded slopes. But on the plateau surface the limestone is often covered by a blanket of boulder clay, and the area is more fertile than the solid rocks would lead one to expect.

It is in its coastline, however, that Gower is most attractive, for almost the whole area from Mumbles Head, at the angle of Swansea Bay, to Worms Head in the extreme west, consists of Mountain Limestone cliffs which vary in form with the changing dip and structure of the rocks. Where the beds are steeply inclined the cliffs are nearly vertical; where they are dipping gently seawards, as is most frequent in eastern Gower, the cliffs are more gently sloping and are there covered with great golden masses of gorse. But these cliffs always lack that smooth regularity so typical of the cliffs of the vale, where the evenly bedded rocks of the Lias have such simplicity. At many places the rocks are broken by faults, which have enabled the sea to cut deep bays which add greatly to the charm of this pretty coast. In the larger bays are high sand-dunes,

and in Oxwich Bay these have dammed up the drainage and produced a great marsh just behind the coast. Still larger expanses of
marsh fringe the northern coast under Llanrhidian and Cheriton,
which stand on an old cliff. In some of the limestone caves remains
of Paleolithic man have been found and Minchin Hole is one of the
most important sites for the study of the fauna which occupied
southern Britain during the Ice Age.

Pembrokeshire

Much of south Pembrokeshire may be compared with Gower.
This 'Little England beyond Wales' is likewise built up of old rocks
sharply folded into east–west anticlines and synclines which have
been planed off by wave action to remarkably level platforms, now
raised to about 200 and 400 feet above the sea. The skyline in south
Pembrokeshire is if anything more even than that in Gower or in
the vale, in spite of the fact that there is a greater variety of rock
types and more complication of structure than in the more eastern
coastal areas. The inland scenery is for the most part subdued,
although the rivers have cut deep but narrow notches into the
generally level surface. Building it up are rocks ranging in age from
Pre-Cambrian to Carboniferous, for here the coalfield is included in
the low coastal plateau, but the surface ignores nearly all these
differences of rock structure and composition, cutting smoothly
across them all (Plate 29). In the north of the country rise the
Prescelly Hills, standing above the coastal plateau of the south as
they do above the Central Wales coastal plateau in the north. Westwards they are represented by isolated rugged hills near St David's
and Fishguard which rise as residuals from the smooth plateau.

The coalfield of Pembrokeshire shows few traces of mining
activity, and its existence scarcely disturbs the pleasantness of this
rich rural county, for no coal is mined there nowadays partly owing
to the complexity of its structure, and the old collieries and refuse
tips rarely obtrude themselves. Yet mining was carried on and coal
exported from Pembrokeshire as early as the sixteenth century, for
the coalfield reaches the coast both at Saundersfoot and in St
Bride's Bay, while the wide estuaries made other mines still more

accessible for shipping. But with the growth of ports farther east, the coal trade of Pembrokeshire has steadily declined, and it remains essentially an agricultural region, green and fertile, with lovely lanes and hedgerows and many pretty villages. ·

Where there is so much variety in the rocks it is not surprising that building materials are also varied in type and in colour. In St David's the older Pre-Cambrian and Cambrian rocks have been drawn upon, and the cathedral, standing in a hollow on the seaward edge of the little town, gains a most unusual colour from the purple Cambrian sandstones. The pale grey Mountain Limestone has been freely used, especially in Tenby and in Pembroke, and Pembroke Castle seems almost to grow out of the rock on which it stands. In this fertile country there are many castles, as there are also in the coastal tracts of Carmarthenshire and Glamorgan.

But if the inland scenery presents the smooth outlines of an uplifted plain, with the different rocks having very little effect on the relief, the coastal scenery is much more varied. Here differences in resistance to wave attack have led to the excavation of bays and to the isolation of numerous stacks and islands, the most famous of which are the 'bird islands' of Skomer and Skokholm. The stretch along the south coast from Tenby around to Angle has much to suggest Gower, for its rocks are chiefly folded Mountain Limestone and Old Red Sandstone, and they give rise to magnificent cliff scenery. The cliffs of steeply dipping limestone in Lydstep and Skrinkle Havens contrast with the red rocks of Manorbier, while limestone cliffs again form much of the south coast around St Gowan's Head and Linney Head, to be replaced once more by red beds near Angle and in the shores of Milford Haven. A belt of volcanic rocks (of Ordovician age, dropped by faulting among the red marls) introduces variety into the cliffs of Marloes Bay, Wooltack and Skomer Island. In St Bride's Bay the Coal Measures have been cut back and form the dark stretch of cliffs from Little Haven northwards to Newgale, while harder and older rocks have resisted erosion along the arms of the bay. The northern arm terminating in St David's Head includes igneous and sedimentary rocks belonging to the Pre-Cambrian and Cambrian, the varied colours of which make up an imposing array of cliffs; the variety of rock types and

Plate 29. *Milford Haven, Pembrokeshire, looking eastwards. The open sea is just to the right of the picture and an oil tanker may be seen in the centre of the Haven heading for the entrance to this natural harbour. The drowned river valley is known as a ria and was probably formed by the post-glacial rise of sea-level. New oil refineries are located on either shore of the Haven. The low plateau-like landscape of south Pembrokeshire is clearly marked.*

colours also leads to a fascinating collection of pebbles in the numerous beaches of the area. The coastal scenery of Pembroke-shire is so imposing that it has been declared a National Park.

One other feature of the coastal scenery may be mentioned, namely the long narrow inlets, especially of Milford Haven and the branches of the river Cleddau; one of these gives access to Haver-fordwest, almost in the middle of the county, where the lowest bridge across the river into west Pembrokeshire has created an important centre. These long narrow inlets are drowned valleys or rias, formed in all probability during the rise of the sea-level which affected southern Britain in post-Glacial times, matching those of south-west England and corresponding in date with the drowning of the Wash area. The same subsidence affected the rest of South Wales, but in some rivers, such as the Tawe, later deposits have filled up the valley to above sea-level. Milford Haven, the largest of the rias, has long been known as a seaport and naval base, but owing to its deep water it has now been utilized as a major oil terminal able to accommodate the largest ocean-going tankers (Plate 29). Thus the coastal landscape of south Pembroke-shire has changed rapidly in a decade, with oil storage tanks, a power station and new access roads engulfing some of the remote rural settlements of the area.

The siting of oil terminals, and to a certain extent the location of nuclear-power stations on the coast owing to their enormous water requirements, are perhaps understandable intrusions into the rural landscape of an industrial society. Less reasonable is the enormous length of the Pembrokeshire coast (within the National Park) which is held by the Ministry of Defence for military training and is therefore closed to the public. Some of the finest coastal scenery is affected, in the Mountain Limestone belt from Stackpole Head to Freshwater West.

19. The Borderland of Wales

Borderlands are always interesting. Where different types of country come into contact, the contrasts of scenic features and of land utilization are often emphasized by a line of market towns where the different kinds of produce are exchanged. The Welsh borders have all these features, but they show besides a wonderful variety of scenery; where the English plain meets the mountains there are numerous castles and border hills of no great height but of surprising daintiness; some parts of the borderland are so fascinating that they call for treatment in this separate chapter. In this, we are not greatly concerned with the present geographical limits of Wales, but propose to describe those transition areas of particular interest which lie immediately to the west of the Midland plain, and more or less along the line of Offa's Dyke.

South Shropshire

Among these, no region is more attractive than that around Church Stretton in south Shropshire. Here a wide range of rocks builds up a scene of extraordinary variety, where the relation of scenery to geological structure is perhaps better displayed than in any other part of Britain. Church Stretton is situated in a narrow and undulating tract of lowland by which the railway passes through the hills between Ludlow and Shrewsbury. It is a pleasant situation, with the Longmynd rising abruptly on the west and Caer Caradoc on the east (Fig. 39). The last-named hill rises steeply from the valley in a face broken only by occasional dark crags, but its eastern face is more irregular, with ribs of rugged rock extending up its flanks. Most of the hill is composed of lavas and ashes produced by volcanoes before Cambrian times, but associated with these are other igneous rocks. The long narrow hill, running roughly north-east to south-west, is cut off by a great fault along its north-western side, a fact which accounts for the smooth face over-looking Church Stretton (Fig. 39). And in line with it are other hills of similar pattern, Ragleth to the south-west, the Lawley to the north-east, similar narrow hog's backs with irregular outlines, and beyond the river Severn, about twelve miles away to the north-east, stands the Wrekin in the same line, rising out still more conspicuously from the plain.

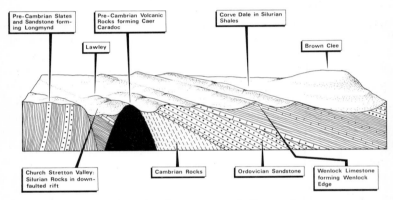

Fig. 39. *The structure of the Church Stretton area, looking north-east.*

Across the narrow faulted trough in which Church Stretton lies is the great bare mass of the Longmynd, rising even higher than Caradoc, but less impressive in its outlines, a smooth-topped plateau stretching for about six miles to the west (Fig. 39). Although near its borders it is deeply cut by narrow V-shaped valleys (of which Cardingmill valley is best known) most of its level upland is unbroken. Large areas on the slopes are covered with bracken, and in autumn they are a rich golden-brown streaked with deep green. Like Caradoc, this plateau area is built up by ancient rocks, belonging to the Pre-Cambrian, but they consist of slates and sandstones producing a mountain of smoother aspect than those formed by volcanic and other igneous rocks in the Caradoc range. The contrast is similar to that previously noted in Wales between the shales and slates of the Denbighshire Moors and Plynlimon on the one hand, and the volcanic mountains of Snowdon and Cader Idris on the other; the materials in these cases are different in age from those just described.

The view from Caer Caradoc eastwards looks first over an undulating tract of fields and woodlands, beyond the Cardington Hill mass (of material comparable with Caradoc itself), into Ape Dale, but the most impressive feature in that direction is the long straight line of Wenlock Edge, a wooded scarp running (as do so many features hereabouts and in Wales) from north-east to south-west, from near Much Wenlock to Craven Arms. This remarkably straight cliff, unbroken for many miles, is formed by the outcrop of a richly fossiliferous limestone of Silurian age, the Wenlock Limestone (Fig. 39). Underlain and overlain by shales, this limestone gives rise to one of the best-known escarpments in Britain, in which the tilt to the south-east results in a distinct and smooth dip slope. Running parallel to this ridge and often less than a mile to the east is another of similar character, made by another limestone, the Aymestry Limestone. Between the two ridges is a rather discontinuous hollow, known as Hope Dale, while the Aymestry Limestone feature in turn dips down to the east and passes into Corve Dale, a beautiful valley connecting Much Wenlock and Ludlow (Fig. 39).

The limestone scarps are often wooded, with bare grey crags. These beds have been (and still are) frequently quarried. The gentler eastern dip slopes, based on the shales, are much more fertile, however, and are cultivated almost up to the crest of the scarp. Both in Corve Dale and in Ape Dale are many small villages, built mostly of the yellow-grey local stones.

The simple structure of the country formed by these Silurian limestones and shales is somewhat modified by changes in the individual scarps, for while the Wenlock Limestone diminishes southwards and forms a comparatively inconspicuous ridge near Craven Arms and Ludlow, the Aymestry Limestone scarp gets stronger in that direction, rising into the fine wooded cliff of Norton Camp and View Edge where it is cut through by the Onny at Craven

Arms. Near Ludlow, moreover, the escarpments lose their straightness, for the rocks are slightly folded and the structure has not the simplicity to be seen in the stretch which is so well seen from Caer Caradoc.

Rising eastwards out of Corve Dale is a great stretch of undulating country, with many fields of heavy red soil marking the outcrop of the Old Red Sandstone. Standing out above it are two flat-topped hills, Brown Clee and Titterstone Clee, which are small outliers of Coal Measures. Brown Clee is capped by a sill of dolerite giving rise to the craggy north-west face which makes the summit a prominent landmark in the country around Ludlow.

Where so many rocks are available there is naturally a great diversity of building materials. Yet old half-timbered buildings are very abundant, notably in Much Wenlock and Ludlow. In the latter town the red tiles, wearing to a deeper tint than the bricks, give the dominant colour, an indication of its situation on the borders of the red marl country (of the Old Red Sandstone). But Ludlow Castle has the yellow-grey of the Silurian stones and many villages nearer Church Stretton also have interesting stone houses. In the area of Soudley and Hope Bowdler, and in the Onny valley near Horderley, much use has been made of a local Ordovician sandstone often showing beautiful purple and yellow-green stripes, quarried in blocks of varied sizes and shapes.

The country west of Longmynd brings us into the actual border of Wales and into scenery which is more suggestive of North Wales, for between Montgomery and Minsterley there are several rugged peaks and much bleak upland. Much of this country is formed by Ordovician rocks which differ greatly from those outcropping around Hope Bowdler, where a series of sandstones gives rise to pleasant rolling country; here to the west of the Longmynd the Ordovician rocks are of Welsh type, chiefly of slates and shales but including beds of volcanic ash and lava. The great difference is probably due to the proximity of the shore-line of the Ordovician sea, coarse sediments being laid down to the east while, in the deeper water to the west, finer muds were laid down in a sea which stretched across Wales and in which there were occasional outbursts of volcanic activity. The harder beds now stand out as ridges, running in a general north-easterly direction, emphasizing the prevailing Caledonian 'grain' of the structures. The most pronounced ridge lies near the western border of the Longmynd, and it is known as the Stiperstones; it is formed by a very hard, light-coloured quartzite, which stands out in bare tor-like crags best known in the Devil's Chair and Cranberry Rock.

Although not the highest mountain in this region, Corndon Hill is peculiar on account of its shape. For while most of the high ground here, as in the Wenlock district, consists of narrow ridges or scarps following the outcrop of the harder bands, with the rivers etching out the softer groups, Corndon Hill is a rounded cone, almost circular in plan, made up of dark grey-green rock which represents a sill of dolerite intruded into the shales. The junction of the resistant dolerite with the readily eroded shales which underlie it is marked by a distinct step in the mountain-side. Corndon forms a real outpost of Wales, and like the Clee Hills, the Longmynd and the Stiperstones, rises to an elevation of about 1,700 feet. The remarkable coincidence of the summit heights suggests that they were carved from a once continuous surface which is now greatly dissected but which can be traced farther west as the High Plateau of Central Wales. From Corndon Hill there is a wonderful view of North and Central Wales, while nearer at hand it overlooks the stretch of exposed moorland around Shelve, where lead mines have scarred the dreary landscape. But very little of the country in south Shropshire has been spoiled, and the varied landscape is nearly always attractive, the hills high and steep enough to be stimulating, detached enough to afford glorious views.

The Country near Shrewsbury

Away to the north of this area the old rocks are submerged beneath the materials of the English plain, and Coal Measures and New Red Sandstone occupy great tracts beyond Shrewsbury and Wellington. Shrewsbury is really a town of the New Red Sandstone, and like Chester it has many beautiful half-timbered houses, besides having many buildings of red sandstone. Its most notable feature is its situation in a great meander of the Severn, so that the old town is almost completely surrounded, and only from the north can entrance be made without crossing a bridge.

The valley of the Severn below Shrewsbury is noteworthy. As far as Buildwas the river flows in a wide, open valley, meandering in the meadows of its flood plain, but then it enters a gorge-like valley where the river flows swiftly for several miles between steep, wooded slopes. The gorge begins when the river crosses the line of Wenlock Edge. The river soon passes on to a tract of Coal Measures and its banks show evidence of early industrial development around Ironbridge. The small tributary valley of Coalbrookdale, in fact, can be regarded as one of the centres of the Industrial Revolution, for here we can still see the forge where Abraham Darby first used coke in the iron-smelting process. Water power was formerly of the utmost importance, but today two large coal-burning power stations dominate the scenery of the Ironbridge gorge.

The change in the character of the Severn valley may have been caused by the Ice Age. It is believed that the Severn formerly flowed northwards across the plains of red rocks to the Dee or Mersey or eastwards to the Trent, but that during the retreat of the ice a stage was reached when the upper Severn drainage was held up by Irish Sea ice invading the Cheshire plain; boulders from Scotland and the Lake District brought by this ice into Shropshire give some indication of its extent. Lakes were held up between the Shropshire hills and the ice front, fed by torrents of water from the melting snows and ice. As the ice retreated several smaller lakes became united, and the water spread over a great tract of country, forming a lake which has been named Lake Lapworth after the geologist who first recognized its significance. Early geologists believed that the waters of this lake rose until they began to overflow by a channel across the original watershed at Ironbridge, where a spill-way was formed and eventually became the Ironbridge gorge, although recent research suggests that the gorge is not an overflow channel but a feature carved out by meltwaters subjected to great pressure beneath a thick ice sheet. Nevertheless the Severn has continued to use and to deepen this channel even after its old course became free of ice, and so its course shows this sharp turn southwards and its valley has the sudden change of character.

The Borders of North Wales

Northwards from Shrewsbury the Welsh borders extend by Oswestry and Ellesmere along the Flintshire coalfield; in most places the administrative border extends far out into the Midland plain, but the geological and scenic boundary lies west of Wrexham and Oswestry where the older, harder rocks rise suddenly from the plain, to be followed in some places by the ancient Offa's Dyke.

Ellesmere, although strictly part of the Midland plain, can be more readily treated here. The meres, which are such a feature of the area, occupy deep hollows with no drainage exits and are surrounded by steep sided hummocky mounds. These are good examples of kettle-holes and kame moraines and are the result of differential melting of a down-wasting ice sheet most of which, in this area, emanated from the neighbouring Welsh mountains. Large blocks of ice, left buried in the sands and gravels, gradually melted causing the ground to subside in the form of enclosed hollows or kettle-holes.

In the Flintshire coalfield the Coal Measures are underlain by the Mountain Limestone which forms a discontinuous but narrow outcrop from Oswestry to the north coast; east of this lies the coalfield with Wrexham and Ruabon, while west of it the older slates and lavas characteristic of North Wales are seen respectively in the subdued contours around Llangollen and in the more rugged Berwyn Hills. The Mountain Limestone rests unconformably on these old rocks, and the country immediately west of it varies in character from place to place, but the limestone at many points stands out in a striking escarpment; in Eglwyseg Mountain the bare white crags make a strong feature facing westwards over Llangollen and the Horseshoe Pass. Near to Llangollen the Dee helps to create a picturesque landscape, where it leaves the mountain land through a series of deep gorges, which in places take the form of large incised meanders (Fig. 40). The main A5 'Holyhead' road, magnificently engineered as gentle gradients through North Wales by the famous civil engineer Thomas Telford, follows the river for some two miles westwards from Llangollen before

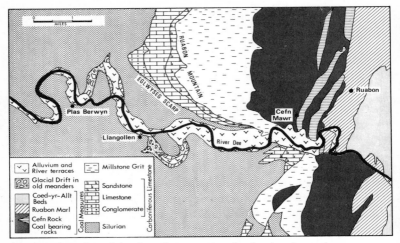

Fig. 40. *The geology of the Llangollen area. Note the abandoned meanders.*

crossing the 'neck' of an incised meander near Plas Berwyn. The river here has not succeeded in cutting through the meander neck although the two abandoned meanders to the north-west and south-east of Llangollen demonstrate how the Dee has already straightened part of its course, probably with the assistance of glacial meltwater streams in the Ice Age. The river obviously pays no attention to the underlying structure and provides another example of a superimposed drainage pattern.

The depth of river incision, where the Dee crosses from the Silurian rocks on to the Millstone Grit and Coal Measures near the small industrial town of Cefn Mawr, has been responsible for another Telford masterpiece; the graceful aqueduct of Pontcysyllte, which carries the canal across the river, was constructed with an attractive local Carboniferous sandstone, the creamy colour of which contrasts sharply with the smoke-blackened, brick-built terraces of the coalfield town across the valley. This excellent building stone, known as the Cefn Rock, from the Middle Coal Measures, has been extensively used in some of the older buildings of the area, with some of the quarries dating back to medieval times. Ruabon church (thirteenth century) and Wrexham parish church are the two most notable examples of its local use, but its employment in the Museum and the Walker Gallery at Liverpool testify to its more general importance and durability. Of even greater renown, however, is the bright-red Ruabon brick which has been extensively used in house building throughout North Wales. It is manufactured from the Ruabon Marls of the Upper Coal Measures, the exploitation of which has left a series of large surface pits and their accompanying brickworks between Wrexham in the north and Pen-y-bont in the south.

The true edge of the mountainland on the borders of North Wales coincides with the outcrop of the Millstone Grit, here known as the Cefn-y-fedw Sandstone. West of the coalfield this forms the uncultivated slopes of Ruabon Mountain (1,648 feet) which dominates the scenery of the Wrexham district, but farther west the dark moorlands and forests of Cyrn-y-Brain rise even higher where an eroded anticline brings Ordovician rocks to the surface. As the edge of the mountainland is traced northwards past Caergwrle it becomes progressively lower in elevation, partly because the Millstone Grit outcrop gets narrower and broken up by faulting. Its place is taken, north of Hope, by Coal Measure shales and sandstones, devoid of coal seams, so that the northern-most borderlands as far as the Dee estuary are characterized by a pleasantly undulating countryside of green fields and coppiced woodlands similar in most respects to the Cheshire plain. But we are reminded that this is still the Welsh borderland when we see at Hawarden the remains of a Norman castle, built like that at Flint, as a base for the English troops when they were engaged in subduing the Welsh of the mountainlands of Snowdonia. The castle and small town of Hawarden stand at the northern end of a low ridge of Millstone Grit and command the extensive marshy plain of the Dee between Chester and the sea. But we have now reached the northern end of the Welsh Marches and must turn south again to Worcestershire where a very different landscape is to be seen.

The Malvern Hills

To the south of Shropshire a most remarkable scenic boundary is found in the Malvern Hills, where a north–south line may be drawn between the red rocks and soils of the New Red Sandstone of the English plain on the east and a series of sharp hills on the west, some of which are a response to the 'windows' or inliers of older, harder Lower Paleozoic rocks which crop out from beneath the Old Red Sandstone of Herefordshire (Plate 30). The Malvern Hills, known from north to south as North Hill, Worcestershire Beacon, Herefordshire Beacon, Hollybush Hill, Raggedstone Hill and Chase End Hill, are built up largely of Pre-Cambrian rocks, but most of them are unlike the Pre-Cambrian rocks of Shropshire; the most frequent type is a gneiss, a granite-like rock in which an irregular banding has been produced by the intense pressure to which it has been subjected. The north–south trend of these Pre-Cambrian rocks also affects the Coal Measures farther north and the Bristol coalfield to the south and may therefore belong to a phase of the Hercynian mountain-building period. Nevertheless, the character of the Malvern rocks is unlike any other outcrop in England and Wales and may represent a slice of Pre-Cambrian basement which is otherwise found at the surface only in north-west Scotland. The eastern face of this range is very steep and may be regarded either as a fault-scarp, the fault in this case resulting from a movement which has pushed up the ancient floor and has brought Pre-Cambrian rocks against the New Red Sandstone, or as a simple overlap of the New Red Sandstone on to the older rocks (as at Charnwood Forest, Leicestershire). Between many of the hills are transverse faults running east–west, along which hollows have been eroded affording routes across the range, such as that between Hollybush and Raggedstone, and that along the Gullet north of Hollybush. These beautiful hills have much resemblance to the ancient hog's backs of Shropshire, and while their heights are little over 1,000 feet they afford extensive views out of all proportion to their elevation, to the east, across the Vale of Gloucester to the Cotswolds, to the west, across the rolling country which extends to Ledbury and Hereford.

Plate 30. *The Malvern Hills, which separate the rolling landscape of Herefordshire, to the left, from the Worcestershire plain seen on the right of the picture. The cols between the higher summits of the Malverns are thought to have been eroded along transverse faults.*

The region immediately west of the narrow Malvern ridge includes a small area of sandstone hills and shale valleys just east of Eastnor, which are based on Cambrian rocks. West of these come the outcrops of Silurian rocks (for as in much of England the Ordovician is unrepresented), consisting of sandstones, limestones and shales comparable with those of Shropshire and forming a similar series of parallel wooded ridges and fertile valleys. Dipping towards the west, these Silurian rocks disappear at Ledbury; and from that town there stretches a great expanse of undulating country of which Hereford is roughly the centre, extending northwards beyond Leominster, westwards by Hay and Brecon, southwards to Monmouth and Newport.

The Country around Hereford

This is the country of the Old Red Sandstone, an area of red soils and rich greens, but with undulating country rising to much greater heights than the areas formed by the New Red Sandstone. Generally these rocks are red, but they include marls and limestones as well as sandstones and conglomerates. Like the rocks of the New Red Sandstone, they probably owe their colour to formation under 'continental' rather than marine conditions, and many beds were deposited in fresh water. The presence of the sandstones and conglomerates explains the occurrence of high uplands in this tract. The most notable feature is the great escarpment of the Black Mountains on the borders of Brecknock and Herefordshire, which may be regarded as an easterly extension of the Brecon Beacons. Like those mountains they rise to over 2,000 feet, the steep northerly scarp rising grandly above the Wye; the gentler dip slope carries them down to Abergavenny, and on this side streams have carved in the upland narrow, parallel valleys much like those of South Wales.

While there are some bare uplands and mountain pastures on the Old Red Sandstone, however, the greater part of the area has a rich soil, and as great areas are under plough, this often adds deep red patches to the landscape. Seasonal changes in colouring are here very marked, for there are extensive orchards and hop gardens, especially in the east, while the masses of daffodils in woods and fields are a delightful feature. Many of the villages and towns are of red brick, but some red sandstone is used and lovely half-timbered houses make many towns extremely attractive: Ledbury, Newent and Hereford itself.

Here and there in this wide expanse of red rocks slight folds have introduced other formations. None of these patches is more fascinating than that at Woolhope, where a somewhat dome-like anticline has brought the Silurian rocks to the surface in an area about four miles south-east of Hereford. The Silurian rocks upraised here consist mainly of limestones and shales like those of Shropshire, and the Wenlock and Aymestry Limestones give rise to scarp and dip slope topography. Owing to the dome-like folding the outcrops are concentric circles, almost complete except along the west, and the escarpments of the limestones form encircling ridges, each with a steep dip slope facing outwards and a scarp facing inwards to the centre. Shut in by this double wall of hills, it is not surprising to find that the area has remained quite isolated, its middle tract having only one village, Woolhope. The main roads skirt the area, that from Hereford to Ledbury keeping just north of the wooded dip slopes which rise so impressively; only minor roads pass through it, following the occasional gaps in the limestone ridges. Its streams, unlike the river Wye, have become adjusted to the structure, tending to follow the shale valleys which carry them round a large part of the area until they escape through gaps such as those at Mordiford and Fownhope to join the Wye.

In the centre of the Woolhope area the oldest Silurian rocks consist of sandstones which form Broadmoor Common and Haugh Wood. The same rocks are brought to the surface again in another fold south-east of Woolhope, where they form the prominent tree-crowned May Hill, nearly 1,000 feet high.

The Forest of Dean and the Wye Valley

Southwards lies the Forest of Dean, the most beautiful coalfield in England. Here the Carboniferous rocks occupy a basin embedded in the Old Red Sandstone, the Carboniferous Limestone forming a rim round most of the coalfield and extending in a broad belt beyond Chepstow (Fig. 41). In structure and relief this coalfield has many resemblances to that of South Wales, for it is partly enclosed by border ridges of conglomerates and Mountain Limestone, and it forms a plateau-like upland in which Pennant

Sandstone crops out extensively. But the differences in scenery between South Wales and the Forest of Dean make these structural resemblances seem unimportant. Here at a height of 700 feet or more, dense oak forests almost hide the collieries and give to Coleford and Cinderford a setting which is unequalled in any other British coalfield.

The best-known scenery of this region is associated with the Wye valley, which between Ross and Chepstow is cut deeply into the western part of the upland. This course is the more surprising when it is recalled that a little farther west there is a much more open tract, between the South Wales upland and the Forest. The river not only takes a course through the high ground, but in its path south of Ross it recrosses the outcrops of Old Red Sandstone, Carboniferous Limestone and Coal Measures several times, and in its great loop at Symond's Yat it leaves the upland altogether only to turn back into it immediately (Fig. 41). The river is thus unrelated both to the present relief and to the geological structure, and its course was apparently determined by other factors, at a time before the Monmouthshire plain had been lowered. Probably the river, with those of South Wales, was first developed on a gently inclined cover of newer rocks, which have subsequently been destroyed, leaving the Wye superimposed on these more complex structures.

The most remarkable features in this part of the Wye valley are the deeply incised meanders similar to those of the Dee at Llangollen, for meanders are characteristic of rivers winding through broad alluvial plains, and are not usually associated with deep gorges. The Wye swings in big curves, some of them with an amplitude of three miles, and in several the river almost forms a complete loop, as in the bend at Symond's Yat already mentioned, and in the great turn under Wynd Cliff, near Chepstow. So the narrow gorge twists and turns as it passes from Ross to the sea, nearly always steep-sided and often wooded, but frequently overhung by precipitous cliffs of Mountain Limestone.

These meanders are presumed to have been initiated when the

Fig. 41. *The geology, landforms and drainage of the Wye Valley and the Forest of Dean.*

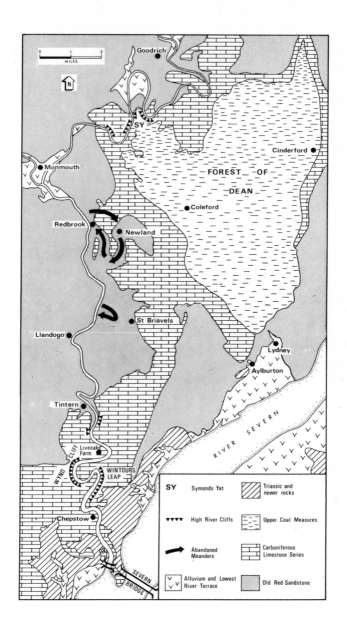

Goodrich

SY

Cinderford

FOREST —OF—

—DEAN—

Monmouth

Coleford

Redbrook

Newland

St Briavels

Llandogo

Lydney

Aylburton

Tintern

RIVER SEVERN

Liveoaks Farm

WYND CLIFF

WINTOURS LEAP

Chepstow

SEVERN BRIDGE

0   1   2
MILES

N

SY   Symonds Yat

▼▼▼▼ High River Cliffs

→   Abandoned
      Meanders

v v   Alluvium and Lowest
v v   River Terrace

Triassic and
newer rocks

Upper Coal Measures

Carboniferous
Limestone Series

Old Red Sandstone

Wye flowed over a wide flood plain in an area of low relief. As the sea-level fell, probably in several stages, the river was enabled to cut down its bed, but there has been no time for valley widening and so its meanders are entrenched. The stages in the evolution of the Wye drainage have been described by Professor Austin Miller who related the incision of the river and the abandonment of the meanders to the various peneplains which he recognized in this region. One of the most striking remnants of these can be seen in the spur on which Liveoaks Farm is situated (some 200 feet above the river) between Tintern and Chepstow (Fig. 41).

The gorge from Monmouth to below Tintern does not wind quite so much as at other points, and its course is followed by the road. The relative straightness of some parts of this stretch, however, is only a secondary feature, for, as in the case of the Dee at Llangollen, there are two cut-off meanders, where the river has shortened its course by finding a way over the narrow neck of the meander. At Redbrook and Newland there must formerly have been a loop like that at Symond's Yat; the old river bed now forms a channel to the east of the present river, parts of the channel being occupied by two small tributaries, the Red Brook and the Valley Brook, while the village of Newland in the dry part of the channel occupies a quite remarkable hanging valley, 'with the church on its very lip'. For the floor of the channel is now nearly 400 feet above the present level of the Wye, a fact which clearly indicates that since this meander was occupied by the river the gorge has been cut down by that amount. Farther downstream, under St Briavel's Castle, is a wide amphitheatre east of the river, and it needs little imagination to realize that the river formerly swung under these cliffs (Fig. 41). But since the level is little above that of the river, it follows that this meander was abandoned much more recently.

Below this, where the entrance to the gorge is guarded by the grey town and castle of Chepstow, the Welsh borderlands may be thought of as ending against the widening Severn. Higher up that estuary, where the Gloucester road follows the low land beneath the slopes of the Forest of Dean, the Old Red Sandstone still influences the scenery, and such villages as Aylburton and others on both sides of Lydney have many brown and red sandstone houses. Newnham stands just on the borders of the Old and New Red Sandstone tracts, on a line which is continuous with the front of the Malverns, but at Westbury-on-Severn the red river cliffs of Keuper Marl and the red brick houses mark the edge of the Midland plain.

20. The Bristol District and the Somerset Plain

On the south-west bank of the Severn estuary and west of the oolite hills is a very varied stretch of undulating country, cut into two by the line of the Mendip Hills. Almost in the centre of the northern part stands Bristol, for long the second city in England, and still rich in old buildings and historical associations. But it is not for this reason that separate consideration is given to this quite small region, nor because the city is surrounded by some of the pleasantest country to be found near any big town, but because the Bristol district exhibits a greater diversity of scenic feature and building materials than is to be found in any corresponding area in England. This is a direct result of its structure and of the great range of rock formations which are exposed within a distance of little over twenty miles, for, excepting only the Ordovician and the Permian, there are outcrops of every main group of rocks from the Cambrian to the Chalk. In this respect, therefore, Bristol occupies a unique situation, and just as England affords more variety in its small area than almost any other country, so this district holds in still smaller miniature many of the chief features of English scenery.

Its villages show such variety of building stones that it would almost be possible to make a geological map of the area merely by noting the materials used in the cottages. Many of the tall-towered churches are built of oolite, though others take their colour from the local stones.

The best viewpoint within the Bristol district is Dundry Hill, a few miles south of the city. This hill has already been noticed as a great outlier of the oolites (p. 25); its slopes are occupied by Lias clays, but the summit is capped by oolitic limestones, which have been used in its cottages and in the church which forms such a prominent landmark. The Lias clays have everywhere proved a poor foundation for the oolitic limestones which form the summit and as a result the slopes of Dundry Hill, like the slopes of the Avon Gorge at Bath, are characterized by masses of foundered rock, most of which slumped during the cold climatic conditions of the Ice Age. This process was not a result of ice-sheet activity but frost-heaving within the Lias clays which stood outside the area affected directly by the glaciers. Landslips have also occurred within post-glacial times, especially on the Lias clays, but not as severely as those which took place when the ice-sheets reached almost to the Bristol Channel. From the northern face of this hill a splendid view of Bristol and the areas east and west of the city is obtained. The even-topped upland of Durdham Down and Clifton is cut by the narrow gorge of the Avon (Fig. 42) and stretches thence westwards in a wooded ridge to Clevedon. From the south side of the hill there is a still wider view of pleasant agricultural country backed by the smooth lines of the Mendip range.

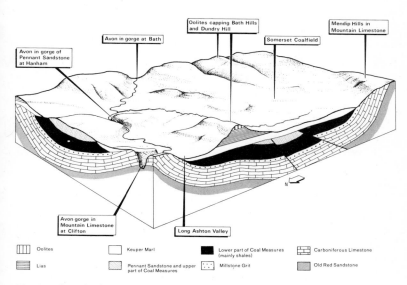

Fig. 42 *The Bristol area.*

Perhaps the most significant feature in the area is the regularity of its hill surfaces, for in any extensive view large areas appear to be flat-topped, whatever their height may be. There are great stretches of country at more or less uniform levels; these are cut by valleys and evidently represent low plateaux in various stages of dissection. But apart from this simplicity of many of its outlines the Bristol district has an astonishing complexity of pattern. Ridges run in almost every direction and rivers seem to flow alternately in wide open valleys and in narrow gorges.

The reason for these features will best be appreciated after an inspection of Figure 42, which gives a very much simplified interpretation of the structure of the area. Two main rock groups are present. The older consists chiefly of the Old Red Sandstone and the Carboniferous, but includes also the Cambrian and Silurian; these have been intensely folded and faulted by a Hercynian mountain-building phase which had culminated before the deposition of the Mesozoic rocks. These newer rocks, including the New Red Sandstone and Jurassic, are of much simpler structure and rest with marked unconformity on the older. The plane of unconformity is not an even plane; it was irregular when the New Red Sandstone began to be laid down, and it has been made still more irregular by such faulting and folding as has occurred since that time. The newer rocks have been removed in many places, and thus the present land-surface appears as a patchwork of different rock types; rivers flow from newer rocks to older and back again to newer, generally carving gorges in the older and occupying wide, shallow valleys in the newer materials, some of which are more easily worn down.

It will be seen from Figure 42 that the higher elevations are of two kinds. There are the Cotswolds and such outliers as Dundry, with a very simple structure essentially depending on a capping of oolitic limestone. Elsewhere there are elevations like the Mendips and the Clifton and Durdham Downs, made up of older rocks, mostly of Carboniferous Limestone, with varied and complicated structures.

The generally east-south-east to west-north-west axes of folding south of Bristol do in fact continue westwards not only into

Gower but also into south Pembrokeshire. All these structures now represent the denuded stumps of a former Hercynian mountain range. It is in part to this phase of folding that we owe the preservation of many of our Coal Measures which, having been downfolded into basins, now form coalfields such as the Bristol, Nailsea and Radstock basins. On the upfolds, or anticlines, the Coal Measures have been completely denuded so that the Mountain Limestone crops out to form the hill masses as at Broadfield Down to the west of Dundry Hill. At most places, however, denudation has proceeded farther, so that the underlying Old Red Sandstone has been revealed at the core of the anticlines. In the Bristol district part of the Old Red Sandstone is known as the Portishead Beds, which form the picturesque coastal ridge between Portishead and Clevedon. The same Old Red Sandstone is responsible for the highest summits of the Mendips, such as Black Down. Nevertheless it is the Mountain Limestone which has given to the Mendips their best-known scenery and most spectacular landforms, and in almost all the places where it occurs it forms high land.

The Mendips

Here the Mountain Limestone gives rise to an area of upland which shows typically those features already described in Chapter 11; there are grey crags above narrow, dry gorges, and rather bare uplands with numerous swallow-holes. Much of the Mendip upland is a limestone plateau recalling that of the Peak District, with fields bounded by grey stone walls.

The Mendip Hills rise from beneath the oolites at Frome and extend westwards to the sea at Weston-super-Mare, where in Brean Down and Worle Hill the steeply dipping limestone forms cliffs which at once recall those of Gower. Out at sea the rocky islet of Steep Holm clearly represents a former extension of the Mendip ridge (Fig. 43).

In many places the Mendip plateau forms a nearly level tract between 800 and 900 feet above sea-level. This is most extensive in the westerly part of the range, between Shepton Mallet and Shipham. In this region there are no villages except Priddy and

Charterhouse, both somewhat sheltered in shallow valleys. Farther east the area reaching this height is smaller, and there are more numerous settlements, while nearer the sea the plateau is more deeply dissected and is represented by such picturesque and almost isolated hills as Rowberrow Tump and Wavering Down.

From this even-topped limestone plateau several low hills rise inconspicuously to about 1,000 feet. Black Down on the north-west is the most prominent of these, its dark heath-covered moorlands forming the gently rounded skyline above Burrington Combe. Three other areas of similar character occur, North Hill (above Priddy), Pen Hill (above Wells) and Beacon Hill (above

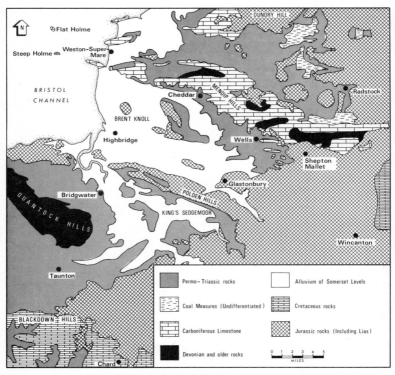

Fig. 43. *The structure of the Mendips and the Somerset Levels area.*

Shepton Mallet). Each of these areas coincides with an outcrop of Old Red Sandstone, exposed in the core of an anticline in the limestone (Fig. 43). The sandstone probably projects above the limestone surface owing to its greater resistance to denudation; in this way these higher parts of the Mendips may be compared with the more prominent hills of Old Red Sandstone which rise above the plateau surface in Gower, where identical structures are found (see Figure 38).

For many people the real charm of the Mendips is to be found along their borders. On the south and west they rise impressively from the fenlands, with a chain of towns and villages occupying the narrow strip between the highlands and marsh, where water supplies are better than on the limestone; Axbridge, Cheddar, Westbury, Wells, Croscombe and Shepton Mallet are all attractive. But the glory of Wells Cathedral, beautiful in itself and in its setting, raises it beyond any comparisons.

Of the gorges which are incised in the Mendip borders, the finest is that of Cheddar, a narrow, dry chasm, cut down hundreds of feet into the upland, exposing great cliffs of grey Mountain Limestone. Evidently it is a valley cut into the hills by running water, but it was left dry when the stream found a course underground, the water now rising to the surface near the end of the gorge. The form of the Cheddar cliffs, steep in some places, sloping in others, is determined by the rock structure. The beds dip towards the south, and as the gorge runs almost from east to west, the northern face is often formed by the sloping bedding planes, while the other face, produced by rock falls along the joint planes, is frequently vertical. The depth and steepness of this gorge, as of all others in the area, owe much to the uplifting of the area in relation to sea-level, which has enabled the stream to cut more rapidly. Ebbor Gorge west of Wells and Burrington Combe on the north of the hills are also narrow defiles making very little impression on the plateau surface.

North of the Mendips the Mountain Limestone again gives rise to Broadfield Down, with its picturesque limestone valleys of Brockley Combe and Goblin Combe, and to the wooded hill ridges between Clevedon and Wraxall, while beyond Bristol a great horseshoe curve of low hills follows the outcrop of the limestone from Alveston to Tytherington and Cromhall and so around the northern end of the coalfield to Wickwar and Chipping Sodbury.

Bristol and Somerset Coalfield

Within the area which is more or less bounded by these irregular limestone uplands lies the Bristol and Somerset coalfield. In many respects this is unlike most other English coalfields, for in the Somerset part the Coal Measure rocks are rarely seen at the surface, since they are concealed beneath nearly horizontal beds of red marl (of the New Red Sandstone) and Jurassic clays and limestones. Much of the Somerset coalfield is a low plateau deeply cut by valleys, a pleasant rural area with occasional black tips along the hillsides but with none of the grimness of most coalfields.

East and north of Bristol, however, the Coal Measures are at the surface over a much wider area (Fig. 42), and the scenery is quite different in character. Included among the Coal Measure rocks is a thick group of sandstones, the Pennant Sandstone, identical with the rocks of that name in South Wales: in the Bristol area the Pennant Sandstone generally forms low hills quite inconspicuous and mostly flat-topped. The contrast between the elevation of the Pennant Sandstone country here and in South Wales is remarkable, for in South Wales this rock group builds much of the plateau which so largely determines the character of that area. The difference is due primarily to the extent to which the plateaux have been raised after their formation, and if the Bristol coalfield were raised 1,000 feet it would probably come to resemble much more closely that of Wales.

A greater similarity is shown by the building stones in the two areas, for where Pennant Sandstone occurs it has been extensively quarried, and many sombre villages, Winterbourne, Iron Acton and Frampton Cotterell, are largely built of the dark sandstone, besides many parts of eastern Bristol, such as Hanham and Fishponds.

The other parts of the area call for no detailed description, for they may be regarded as forming an extension of the Midland plain. Thus red marls fill many valleys and lap around the lower slopes of many of the uplands, and they also underlie much of the fen at various places along the estuary, where they have yielded rapidly to the effects of denudation. To some extent it may be said that parts of the area represent a fossil landscape, comparable with Charnwood Forest in that denudation tends to reveal the form of the older uplands as the softer marls are more quickly worn away.

Besides the red marls, the clays and limestones of the Lias occupy considerable areas, where the cottages of small blue and yellowish stones form a constant feature, especially noteworthy around Keynsham, Pucklechurch and Radstock.

The Bristol Avon and Its Tributaries

In this rapid survey of the area no reference has yet been made to the rivers. The Avon gorge is the most impressive scenic feature in the immediate neighbourhood of Bristol, by which the river passes from the low ground south of the city through the limestone upland of Durdham Down and Leigh Woods; in the gorge the sheer cliffs are formed by beds dipping steeply towards the south, Mountain Limestone for most of its length, Old Red Sandstone at the northern end (Fig. 42). The origin of the gorge can only be appreciated when something of the history of the region is understood, for at first sight it appears surprising that the river does not keep to the left along the easy path which appears open to it through the valley at Long Ashton (Fig. 42). That it has kept to the course through the gorge, cutting its way down through hard limestone, does not seem so extraordinary when it is realized that the alternative valley probably did not exist when the river commenced to cut its gorge, and that the low district south of Bristol then stood rather higher than Durdham Downs. In short, while an observer is likely to be most impressed by the work done by the Avon in cutting its gorge, the river has actually moved much less material here than elsewhere in its course. Weathering has done little to wear down the limestone above the gorge, whereas it has much more rapidly attacked the readily eroded clays and marls, and has helped in the excavation of a wide valley at Keynsham. The Avon gorge, like most other great gorges, is due to a river which at that point has had little energy to spare for anything but downward cutting.

To understand fully the Avon gorge and its relations to the river scenery of the rest of the area it is necessary to go further back along the valley, and to follow the river from Bath, where it leaves the lovely valley between the oolite hills and flows across the wide alluvial plain beyond Keynsham to enter another gorge at Hanham (Fig. 42). This is less beautiful perhaps than that at Clifton, because it is cut into the darker rocks of the Pennant Sandstone, but the sides are often wooded and the scenery is by no means unimpressive, particularly towards Conham, where the river sweeps round a deeply cut meander. From this point the Avon crosses the lower ground through Bristol before entering its better-known gorge. After leaving Bath it thus flows alternately in open valleys and in narrow gorges cut into uplands of varying character and height. In these respects the Avon recalls the Wye, and it is highly probable that the later history of these two rivers has been similar. The lower parts of both river basins had probably been worn down to a fairly level plain before an elevation of the whole region (amounting to several hundred feet) gave additional cutting power to the rivers and enabled them to carve their gorges.

The rivers of the Bristol district, however, illustrate some features of their development more clearly than does the Wye system. For whereas the Wye, like the rivers of South Wales and of the Lake District, is believed to have been initiated on a cover of newer rocks (now totally disappeared) and to have become superimposed on the present structures when the newer rocks were cut through, in the Avon and some of its tributaries we see a river system just in the act of becoming superimposed. In this respect it is one of the most fascinating river systems in Britain. Probably rivers began to flow across the area when it was blanketed under a thick cover of gently dipping Mesozoic beds, including the Chalk. Some believe that Welsh rivers may once have flowed across the present position of the Bristol Channel in a direction opposite to that of the lower Avon, possibly linking up with the river Kennet and thus with the lower Thames. Evidence for this is seen in the easterly flowing headwaters of the Avon on the Jurassic rocks near Malmesbury, and the left-bank tributaries of the Avon which cross the Radstock coalfield from south-west to north-east, before turning sharply westwards when they join the Avon gorge between Bradford-on-Avon and Bath. Whether or not this is a true reconstruction, we can be certain that rivers cutting down first through the Chalk, and then the Jurassic and New Red Sandstone, met older rocks more quickly in those places where they rose in elevations and the cover was thinnest than in others. Some streams have only just cut through this cover and reached the older rocks beneath, which thus appear only along parts of the valley floor, as in Vallis Vale and other valleys near Frome; elsewhere rivers have not yet cut completely through the newer rocks.

Some of the tributary streams show features very closely resembling those of the Avon. The Frome has attractive gorges where it cuts into the Pennant Sandstone at Stapleton, Frenchay and Winterbourne. The Chew and its tributaries are likewise entrenched near Pensford, while the Trym streams occupy astonishingly deep valleys cut in the Mountain Limestone near Westbury-on-Trym. It is clear that all these features are of similar origin, and it is fairly certain that all are to be attributed to the partial superimposing of the river system on a region of complex structure.

A final question has yet to be answered. Why has the Avon drainage system made an apparent change of direction from an easterly flowing stream to one which now enters the Severn estuary? The answer seems to lie in the structural history of the Bristol Channel, which some authorities now believe is a relatively late phenomenon following a period of trough-faulting in Tertiary times. This rapid change of base-level would certainly encourage a river to cut rapidly down towards the west, leaving the east-flowing drainage partly dismembered.

The Somerset Plain

To turn briefly to that part of the Somerset plain which lies south of the Mendips, we may notice that it extends westwards and southwards to the foot of the Quantock Hills and beyond in a narrow belt to Minehead and the borders of Exmoor. Much of this country is underlain by the red Keuper Marls, which here as always form land of low elevation. Included in the area are the great flats of the Somerset 'levels': the fenlands of the west where silt and peat now fill ancient valleys (Fig. 43). In these wide green marshes, crossed by the main road from Bristol to Bridgwater, habitations are few and are almost confined to the borders of the willow-fringed roads which are raised slightly above the field level; in many stretches of country the greatest heights are reached where bridges cross the drainage canals.

The few islands which rise out of these marshes are thus un-usually conspicuous, and have been especially attractive to settle-ments since earliest times. The most famous among them is the lovely Isle of Avalon with the town and ruins of Glastonbury. This rises to a lofty hill from which the most wonderful view of these fens is to be had (Plate 31). Farther west Brent Knoll has an even greater isolation, overlooking the coast from Brean Down to Burnham and beyond. Both of these islands are really outliers of Lias limestones and shales, which formerly rose out of the gentle valley whose submergence has formed the marshes. The nearly horizontal beds are traceable in the form of the hills, each harder band forming a terrace on the hillsides. The Isle of Wedmore is larger, and like the hills south of Wells it is partly built of red marl, although Lias limestones form the western part. The origin of the Somerset levels has caused a great deal of speculation, for although their post-glacial chronology is fairly well described, including the marine transgressions of Neolithic and Romano-British times, little was known of their history during the Ice Age. It is now known that one of the earlier ice sheets which moved down the floor of the Irish Sea basin impinged on the northern coast of Devon west of Lynmouth, thus impounding the entire drainage of the Severn estuary which lay outside the glacial limits. A large pro-glacial lake may have been formed over the Somerset plain and it has been suggested that it overflowed through a gap at Chard before escaping to the English Channel via the valley of the Seaton Axe. The rem-nants of this speculative lake are thought to have been the fore-runners of the Somerset levels.

Westwards the red marl country may be traced to the Vale of

Plate 31. *The town of Glastonbury with its famous 'Tor' beyond. Glastonbury itself is situated on a shelf of Lower Lias clay which stands above the marshes of the Somerset Levels. The hill on which the tower stands, known as the Tor, is an outlier of Upper Lias clays and sands, the main outcrop of which can be seen in the Corton Hills and Cadbury Castle in the far distance.*

Taunton Deane, the rich agricultural region surrounding Taunton and Wellington, where villages and farms are more numerous and more evenly distributed than in the fenlands. The same red country extends by the Quantocks near the coast, but for many miles the cliffs are formed of Lias shales and limestones, which around Kilve and Watchet give rise to low, blue-grey cliffs, much like those on the opposite coast of South Wales, where the similar rocks in the Vale of Glamorgan emphasize again the former continuity of these regions.

21. Devon and Cornwall

The only great tract of England which has not been described lies west of Exeter, and includes Cornwall, most of Devon and a small part of Somerset. It is a remarkably varied region, best known perhaps along its coast. The cliffs of the north are high and scarcely broken for miles, but although the south coast is quite rocky, it is not so steep and is broken by innumerable narrow inlets of great beauty. Inland there are lovely lanes between tall hedges, rich, green country rising to billowing hills and to the open moorlands which occupy thousands of acres on Exmoor and Dartmoor.

The main impression of this area is of rolling plateaux at various heights, dissected by deep river valleys, and fringed by a most beautiful and varied coast. Some of the region is fertile in spite of its altitude, and with its comparatively mild climate, farming is carried to high altitudes.

The border of this region, while not coinciding with county boundaries, is clearly marked. It is the line between Red Devon, made of New Red Sandstone, and the generally brown and grey older rocks of the peninsula to the west. We may think of the New Red Sandstone as belonging to the English scarplands of lowland Britain, while the rest of Devon and Cornwall essentially belongs to highland Britain, although the highlands in this case have been much worn down with the passage of time. Many travellers to the west will cross this marked boundary near Exeter. From there it runs nearly due south to near Torquay, while to the north it passes by Cullompton and Milverton to trace the margin of the Brendon Hills to the coast near Minehead (see geological map, back cover). It is only a little obscure where a finger of Red Devon points westwards through Crediton and North Tawton, where the Blackdown

Hills of the English scarplands dominate the older rocks to the west, and where the Quantocks stand out from the lowland near Taunton as a scenic extension of north Devon.

At this boundary there is a great unconformity, the old rocks passing under the New Red Sandstone and younger rocks. These older rocks chiefly belong to the Devonian and Carboniferous, and have been intensely compressed and folded with an east–west trend before the young rocks were laid down. The intensely folded rocks are arranged in a broadly synclinal structure stretching east and west, with the Carboniferous rocks occupying the central area, and older Devonian rocks to north and south. Thus at the boundary the whole 'grain' of the country alters, as well as the rock type and age, from the east–west trend just discussed to the south-west to north–east trend of the English scarplands (although complicated by the overstep of the Upper Greensand and Chalk over the Jurassic and New Red Sandstone rocks) (pp. 69 and 125).

Exmoor and North Devon

The northern part of the peninsula may best be described by imagining an approach. such as many travellers now make. through Bridgwater. Beyond that town the red marls and clays give place to Devonian slates which rise suddenly to over 1,000 feet in the narrow ridge of the Quantock Hills. The western edge is a steep and wooded fault-scarp, but the north and east is deeply cut by beautiful combes carved by swiftly flowing streams rushing to the coast. Small villages are situated in some of these valleys, but the most important settlements are found where the hills meet the plain.

If the Quantocks are to be thought of as a great detached outlier of the north Devon moors, the hills between Minehead and Porlock may be regarded as a similar but smaller and less isolated mass, on which Dunster Castle forms a prominent landmark. The low-lying valley behind Porlock is underlain by the New Red Sandstone and separates Selworthy Beacon from the great tract of Exmoor. Faulting is responsible for these features. Hereabouts the red fields and brick cottages recall the Midlands, but the famous hill out of

Porlock carries the road up to the summit of the moors, well over 1,000 feet high. Thence through Lynmouth to beyond Ilfracombe is one of the most impressive stretches of cliffs in England, falling almost sheer from the moorland plateau to sea-level. It is impossible to think of Exmoor apart from this line of cliffs and the incomparable views afforded across the Bristol Channel. It is possible that the cliffs are in part a fault-scarp, and that the Bristol Channel was formed by faulting in Tertiary times (comparatively recently, geologically speaking). Thus these magnificent cliffs have survived in their present form, with the moorland surface just behind them about 1,000 feet above sea-level, rising inland. The proximity of the coast gives character to the moor; the sharp change of level gives energy to those moorland streams which tumble northwards into the sea, carving dark and narrow valleys into the otherwise smooth upland.

Exmoor extends for over thirty miles from east to west and is nearly twenty miles across at its widest part. It is an area of smoothly moulded hills with little bare rock showing, the dull green plateau much resembling parts of central Wales. Reaching a height of some 1,400 feet, it preserves this level over great areas, but out of it rises Dunkery Beacon as a residual some 300 feet higher. This plateau, like that of Wales, cuts across beds of various types and it must be the result of either sub-aerial or marine action, which planed the area down to near the sea-level of that time. Research has in fact shown that Exmoor is not just one simple surface with residuals, but several surfaces in a subdued step-like arrangement. The present height of these surfaces has been determined by subsequent uplift, but of their history there is as yet little known. They may be comparable in age with the Middle or Low Peneplains of central Wales and perhaps, like them, were raised to near their present position in late Tertiary times, but they must not be regarded simply as an extension of the Welsh upland. The Bristol Channel lies between the two, and as we have seen, this is probably faulted, and thus we cannot expect any exact coincidence of levels across the channel. However, this is all very speculative and much more evidence is required before these aspects of Devon scenery can profitably be discussed.

Plate 32. *Lynmouth, Devon. This village lies at the mouth of the East and West Lyn rivers draining from Exmoor. Some idea of the steep cliffs and nature of the relief of the area can be gained from this photograph. The present controlled river channel is in the middle of the photograph. The flat land of the village is the 'delta' at the river mouth, built of the debris from former floods, not least that of 1952.*

On Exmoor there are fields up to over 1,000 feet, but much of the higher land is rough hill pasture, often very bare and desolate, bracken-covered and sometimes boggy. As in many parts of Devon, the fences are often banked on a rough wall of slates. In the sheltered, wider valleys, narrow, green fields run up the slopes, but the narrower valleys (especially those draining to the north coast) are heavily wooded. The valleys shelter scattered habitations; colour-washed farmsteads with slate roofs are fairly abundant, and the moorlands hold some villages, of which Simonsbath in the valley of the Barle is the most central. But in any wide view of the plateau, valleys are inconspicuous and houses are lost to sight.

The coastal border of the moor is so abrupt that there is no room for villages near the shore on the valley slopes. Lynmouth, the only place of importance on the shore in a stretch of over thirty miles, has barely space to stretch by the road or riverside, for the picturesque wooded valleys have almost precipitous sides (Plate 32). Lynmouth's position at the mouth of the deep valleys of the East and West Lyn rivers proved disastrous in August 1952, when a tremendous flood swept through the village, causing extensive damage and loss of life. The moors had been saturated by heavy rain early in the month, and so the waters from a torrential storm swept down the valleys and through the village, bringing vast quantities of boulders and debris and diverting a river course. Such catastrophic events often bring about more change almost overnight than decades or centuries of normal erosion. Combe Martin is similarly strung out for over a mile along the deep valley of the Umber. The growth of Ilfracombe has been controlled by the relief.

In the high cliffs the rock structure of the area is well displayed, and the diversity of rock type shows itself in the varied form of the cliffs. For although the rocks dip generally to the south, the trend or strike of the beds is not quite parallel to the coast, and newer beds form the cliffs as we pass westwards. Thus at Lynton massive sandstones form the cliff at Castle Rocks, the beds and joint planes weathering out in rectangular blocks suggestive of a ruin. These bare rocks form an impressive bleak cliff, while in places farther west softer slates have been worn back to form small bays where trees creep down to sea-level, as in Woody (Wooda) Bay. Where the slates form projecting crags and stacks they have a ragged appearance quite unlike that of the bolder sandstones. The heights of the Great and Little Hangman, near Combe Martin, are due to sandstones which project as headlands owing to their greater resistance to wave advance. Along much of the coast grey slates form the irregular cliffs, but at several points the rocks are red, and the varied colours of the crags combined with the rich green of the vegetation make a coastline of great beauty.

These rocks belong to the Devonian system; they were formed mainly as deposits in a shallow sea but they have been greatly influenced by subsequent earth pressures, which altered the muddy sediments into slates and contorted many beds. They are contemporaneous with the Old Red Sandstones of South Wales, and while it is thought that the latter were laid down in fresh water, the redness of several groups on the north Devon coast may indicate that these two regions were not completely separated at all stages while the rocks were deposited.

The coast west of Ilfracombe is quite different in character. Here the shore-line runs across the rock groups almost at right angles, and the harder beds project in more striking headlands than on the coast to the east, while the softer rocks have been more rapidly cut down. The contrast between these two stretches of coast is thus similar to that pointed out between the east of the Isle of Purbeck and the coast at Lulworth. At Woolacombe soft beds have been eroded to form Morte Bay between Morte Point and Baggy Point, which, like the smaller rough crags under Mortehoe, project westwards with the trend of the harder rocks: Croyde Bay again picks out a softer belt.

In Barnstaple Bay the shores are often low. The sand-dunes of Braunton Burrows, blown inland by westerly winds and banked at Saunton against the dark cliffs, have led to the formation of extensive marshes 'where Torridge joins her sister Taw, and both together flow quietly toward the broad surges of the bar'. By these sheltered estuaries the picturesque towns, Bideford, Barnstaple, Appledore and Instow, are alive with memories. They were once prosperous ports for sailing ships, but the estuary is too shallow and tortuous for all but the smallest modern ships, and the flat land has been used for a naval air station, whose jet aircraft now scream across the bay.

Most of Barnstaple Bay is cut into the Carboniferous rocks, for the Devonian rocks which occupy north Devon dip southwards, under the Carboniferous, near Barnstaple. But the Carboniferous rocks are not like those of any other part of England : there is little limestone, and no tract can be likened to the Mountain Limestone areas across the Bristol Channel. The higher parts corresponding to the Coal Measures contain no important coals, but beds of culm, a soft powdery coal, occur occasionally. The Carboniferous rocks consist mainly of dark shales and sandstone, intensely crushed and folded. The boundary between the tough Devonian slates and sandstones of Exmoor and the softer Carboniferous Culm Measures is clearly marked topographically in an east–west line from Barnstaple to Bampton and Milverton. To the south of this boundary the Culm Measure country is lower, in a series of levels generally between about 800 and 500 feet above sea-level. These plateau surfaces have been deeply cut by the river valleys, whose peculiar features will be discussed presently.

These Carboniferous rocks occupy the cliffs southwards to Bos-castle in north Cornwall, for the most part a bleak and rugged coast, characteristically dark grey or brown in colour. The cliffs are never dull, however. They have such diversity of form, their dark colours react to every change in sky or sunlight, and vegetation brightens every ledge.

Perhaps this coast is best known at Clovelly, on the south of Barnstaple Bay, where cliffs several hundred feet high form the clean-cut boundary of an undulating tract of varied fertility. As in so much of north Devon, there is little low ground below the cliffs and there are no natural harbours, and Clovelly gains in picturesqueness from its position down the steep slope of the cliff. Along the Clovelly coast to Hartland Point the cliffs run nearly with the strike of the rocks, but from that point southwards the coast again cuts across the bedding almost at right angles. The coast of north Devon and Cornwall thus falls into four sections : that from Porlock to beyond Ilfracombe and that near Clovelly having cliffs which almost follow the trend of the rocks from east

Plate 33. *Intensely folded rocks near Boscastle Harbour, Cornwall. Folds have been refolded here. The rocks are black slates, siltstones and laminated sandstones.*

to west, that from Morte Point to Westward Ho! and that from Hartland continuing to Boscastle in Cornwall having the cliffs cutting across the rock groups. The latter show the greatest diversity of pattern and the finest rock scenery. From Hartland to near Bude the harder rocks form ribs which project out at right angles from the shore, giving rise to innumerable small, rough headlands, the shapes depending on the dip which varies with every fold in the rocks. In places rock ribs have been cut through along their joint planes, and portions isolated as stacks, while in the softer beds caverns have been excavated. The intense folding and crushing, too, can plainly be seen in section in these cliffs (Plate 33). It is a coast which is slowly receding under the wave attack.

In places streams are carving deep valleys, usually too narrow to hold a road but beautifully wooded and in summer full of wild flowers. Many of these little streams, which drain the land for only a few miles from the coast, form most attractive waterfalls where they reach the cliffs. The late Dr E. A. Newell Arber made a careful study of these coastal falls, and showed how they result from the coast being worn away. At its mouth a stream will always tend to carve down its valley to sea-level, and will 'grade' its bed to that level. If, however, the coast at its mouth is being worn away more quickly than the river cuts downwards, then the river does not reach sea-level at its mouth and the valley is left 'hanging', the stream falling over coastal cliffs.

Such coastal waterfalls are perhaps better seen in the coast near Hartland than in any other part of England. Various types may be found. Litter Water, some nine miles south of Hartland Quay, ends in a sheer fall of seventy feet; seen from the beach it is apparent that the stream has carved a valley which makes a notch on the cliff-top, but the valley is obviously hanging high above the shore, the almost vertical cliff being formed here by steeply tilted beds. In other cases the streams cascade over more irregular cliffs, especially where beds of different hardness have caused steps in the descent.

Some valleys at the cliff edge are now, however, thought to have been caused by the proximity of an ice sheet. It is now known that one of the advances during the Ice Age caused an ice sheet to reach the north coasts of Devon and Cornwall in part at least, possibly crossing it in places, and stretching as far as the Scilly Isles. Deposits at Stony Porth in the Scilly Isles, at Trebetherick on the Camel estuary, at Fremington on the Taw estuary, and in Croyde Bay have been identified as glacial till, and giant erratic boulders have been recorded at Croyde and Saunton on the north coast of Devon, and also at Porthleven on the south coast of Cornwall in Mount's Bay. The latter is a fifty-ton block of garnetiferous gneiss; there is still controversy as to whether this boulder would require an ice sheet to emplace it, or whether it could have been moved into position embedded in an iceberg or even pack ice.

The ice sheet that deposited the till must have pressed against the high-cliffed coast at least between Lynton and Hartland Point, and may have produced temporary drainage diversions, forming channels running along the margin of the ice. Possible marginal channels have been identified at the Valley of the Rocks, west of Lynton, and near Clovelly and Hartland Point. Previously these channels have been interpreted as resulting from river capture and rapid cliff recession. This new explanation may necessitate a revision of the estimated rate of cliff recession.

At the time of this ice advance, and possibly at other periods in the Ice Age, Devon and Cornwall, together with the rest of unglaciated southern England, endured a periglacial climate, with many effects which can be seen in the landscape today. The most widespread are deposits of 'head', a mixture of clayey material and frost-shattered stones which was able to sludge down quite gentle slopes. This process causes some rounding-off and lowering of slopes throughout southern England: its effects have already been noticed in south-east England (p. 59). More spectacular results of a periglacial climate can be seen in the granite areas; these are discussed below.

The Rivers of Devon and Cornwall

Most of the lovely rivers of this area have some most unusual features, and there is no generally agreed explanation of these peculiarities yet available. Nearly all are deeply incised into the plateau surfaces, and most have picturesque estuaries.

Many of the major rivers of the area rise near the north coast, and those debouching to the north almost all have peculiar courses. This is consistent with the idea that the drainage began as a series of streams flowing north–south across the area from sources near the north coast, rather as the Exe still does today, rising high on Exmoor, near the short, dark, deep valleys which drain north to Lynmouth. From its source the Exe flows almost due south across a variety of rocks to Exeter and Exmouth, with a fine disregard for structure.

We may consider by contrast the Taw and Torridge which flow to their common estuary in Barnstaple Bay. The Taw rises on Dartmoor and flows north across the complex structures, receiving on its way the Bray, which drains part of Exmoor, flowing south-wards. The waters of the Bray almost double back on themselves when they join the Taw. The Torridge, meanwhile, rises within a few miles of the sea near Clovelly, and flows in a south-easterly direction for about twenty miles before almost doubling back on itself to flow west of north to its present estuary. There is comparatively low ground dividing the Exe, Taw and Torridge and their tributaries in their middle courses: this low land is the finger of Red Devon pointing westwards which was referred to in the opening pages of this chapter. There are grounds for thinking that the Exe has lost the Taw–Torridge systems because of the advantageous shorter, steeper course to the north coast.

Such patterns as these are repeated in the river systems to the west. The Tamar rises near Hartland Point and flows south, but it is in imminent danger of losing some of its headwaters to the short river Bude, which reaches the sea at Bude Haven and has already captured the river Strat from the Tamar. The Camel is rather similar to the Torridge; it rises near Boscastle on the north coast and flows south, until near Bodmin and the river Fowey it turns and flows north-westwards to its estuary at Padstow, leaving a wind gap near Bodmin between the Camel and Fowey.

It is thought that the original south-flowing streams with their sources near what is now the north coast must have originated on a planed surface which tilted southwards, and from which rose the highlands of Exmoor, and Dartmoor and other granite masses. It is uncertain whether this was a marine surface which was formed in early Tertiary times, traces of which remain around the highlands and in east Devon on Upper Greensand (p. 69). or a late Tertiary sub-aerial surface which can be traced lower down the same slopes and on into the summit planes of the Cotswolds and the chalklands of south-east England. The subsequent history of Devon and Cornwall has seen the reduction of this surface and the highlands, by a series of stages, to the form we see today with its diverted and incised rivers, and the magnificent drowned estuaries. This later history will be discussed below.

Dartmoor and Other Granite Moors

We must now turn our attention to the southern half of Devon and Cornwall, and the most striking feature of this area, apart from the coast, is probably the wild moorland of Dartmoor. It is an undulating country some 200 square miles in area, reaching up to 2,000 feet. Although the moor is crossed by two main roads, most of it is trackless, and can be seen only on foot or horseback. Much of the area is bleak and treeless, except in the valleys; there are dark peat tracts above which the slopes, at times purple with heather or golden with gorse, rise to bare grey 'tors', weathered into queer shapes often resembling tumbled ruins. Dartmoor is wilder than Exmoor, showing more bare rock and many 'clitters' of great boulders (Plate 34).

Dartmoor differs from Exmoor primarily because it is made of different rock: while Exmoor is of Devonian sandstones and slates, that is sedimentary and metamorphic rocks, Dartmoor is made of igneous rocks, formed by the consolidation of molten material. Granite is the rock of the moor, grey in colour and coarse in grain. An enormous dome-shaped mass or boss was forced upwards after the Carboniferous rocks were formed, during the intense folding. This mass cooled and crystallized at some depth below the surface, and has been exposed at the surface much more recently, when its cover was stripped off by weathering. The result of this cooling at depth is a typical crystalline rock, consisting of white crystals of feldspar with black and white mica and glass-like quartz. The granite is generally harder than the surrounding rocks, so that it projects above them. Around the granite itself, some of the sedimentary rocks into which it was intruded were affected by the immense heat and pressure, and have been metamorphosed in an 'aureole' around the main granite mass.

Since the granite has been exposed to denudation it has been reduced to a number of surfaces representing higher base-levels. An upper surface at between 1,900 and 1,500 feet, sloping southwards, has been recognized, from which rise the residual areas reaching over 2,000 feet. There is another well-developed surface between

Plate 34. *Dartmoor. East Mill Tor, about three miles south of Okehampton, is in the foreground. This is a typical granite tor, showing horizontal and vertical 'pseudo-jointing'. The view looks west, across the valley of the Black-a-ven Brook to the granite moor rising to Yes Tor, 2,027 feet above sea-level and just over a mile away.*

about 1,350 and 1,050 feet, again tilting to the south. This could possibly be the surface on which the southerly drainage of much of Devon and Cornwall was initiated. A further lower surface occurs all around Dartmoor at between 950 and 750 feet and is fairly level. This is thought to correspond with the late-Tertiary summit surfaces of the Cotswolds and the chalklands. The river valleys exhibit a series of lower terraces.

   The granite consists of a vast thickness of almost homogeneous material, extending downwards for unknown depths into the earth's crust, but nevertheless it is cut by more or less regular sets of 'pseudo-joints' (an igneous rock cannot have true joints and bedding), along which it becomes worn into blocks. Such ready-made blocks of 'moorstone' have long been gathered for building materials, and granite is the essential building stone over a wide area. Yielding great rectangular blocks, it has been used in building simple and solid churches and houses. Little decoration is possible in these granite buildings, for the material is not easily cut, but it expresses admirably the grimness of the moor.

A characteristic of these granites is their tendency to break down into their individual crystals under the attack of weathering agents, particularly when the climate is warm. Weathering may take place more readily and deeply along the line of these pseudo-joints, leaving blocks of sound rock amidst deeply weathered rotten granite. An example of this can be seen in section in a quarry near Two Bridges. If the rotten material were to be removed, then the blocks might be left standing, and this is a probable explanation of the tors and clitters found on the granite moors. The agencies for the removal of material have probably been most active during the periods of periglacial climate of the Ice Age. It is thought that some feet, or even tens of feet, of material may have been removed from the granite highlands at these times, leaving the tors and clitters as remnants, with some modifications to slope forms.

Several rivers flow from Dartmoor – the Plym, Tavy, Teign and Dart. The moor was a highland at the time when the southerly drainage described above was initiated, and has remained so, with the result that it has its own rivers radiating from it. These rivers at first occupy wide valleys on the uplands, but before they reach the granite border they plunge into deep wooded gorges. The main road from Ashburton to Tavistock crosses the Dart in this gorge stage at Holne Bridge, New Bridge and Dartmeet Bridge. Many of these gorges are incised in the floors of wider, older valleys, marking stages in the uplifting of the region. Some reservoirs for water supply are found around the moor.

Numerous villages and a few towns are situated near the border and in the valleys just within the moor, Ivybridge, Shaugh Prior, Tavistock, Mary Tavy, Chagford, Widecombe-in-the-Moor and the rest, but there are few places of any size on the moor itself. Princetown is well up on Dartmoor, a grey, bleak town 1,400 feet above sea-level, but otherwise there are few habitations except isolated farms. Yet Dartmoor was for long a home of early man, when dense forest made much of the lower land unsuitable, while the moor itself was more lightly forested. Ancient stone trackways, avenues of standing stones, stone circles, barrows, hut circles, all tell of early occupants. Forests are now reappearing around some margins of the moor in the form of Forestry Commission plantations.

Dartmoor is the largest and best known of five similar large granite masses or bosses. The other four are, from east to west, the Bodmin, St Austell, Carn Menellis and Land's End masses, while the Scilly Isles represent the surviving remnants of a sixth. Each of the five gives rise to moorland rather similar to, but lower than, that of Dartmoor, and forms an area of higher ground rising above the surrounding country.

Bodmin Moor is the highest point of Cornwall, but comparatively little of it stands above 1,000 feet. From many of its western or southern hills there are splendid views across Cornwall, for the regular coastal plateau stretches for miles, embracing the whole of the peninsula save where the low granite hills stand out. Of these, seen from Bodmin, Henbarrow Down (part of the St Austell granite mass) is most conspicuous although it only just exceeds 1,000 feet.

The relief of Bodmin Moor is not very closely related to the drainage, for like Dartmoor this area is cut by two wide shelves or plateaux at heights of about 800 feet and 1,000 feet above sea-level. The higher of these plateaux is seen only in isolated patches on the upper parts of the moor, but there can be little doubt that these were once connected and that the valleys have been cut down subsequently. Seen from the north, from a viewpoint on Davidstow Moor, the regularity of this surface is apparent, while Brown Willy and the higher tors stand out conspicuously, and obviously represent ancient residuals. The lower shelf is more extensive, and is easily recognized around Camelford.

The St Austell granite is much less extensive, and although it stands out from the surrounding slate areas, it has few rugged tracts and few tors. Over much of the area there are low rounded boulders, sometimes over thirty feet in length, the weathered relics of former tors. The appearance of this upland has been greatly modified by the huge clay pits and dazzling white dumps of waste resulting from the china-clay industry. China-clay rock is an altered form of granite, and although some occurs on Dartmoor and on Bodmin Moor, the St Austell district (especially the western part) has long been famed for the purity of its china clay. Although it was used for pottery as early as the eighteenth century, it has only been worked on a large scale during the past few generations. China clay rock has been derived from normal granite by a process of 'kaolinization', that is by the chemical alteration of its crystalline feldspar into kaolin, a white clay mineral. When this alteration is complete, the other minerals present in the granite (quartz and mica) can readily be separated. There has been much discussion as to the cause of kaolinization; somewhat comparable changes take place in any granite as a result of weathering agents, but the great extent of the alteration of the Cornish granites, the arrangement of the altered material in parallel belts associated with fissures and the relation of these to tin veins, suggest that the origin of the china clay is connected with peculiar chemical conditions which existed during the cooling and consolidation of the granite. At that time rising vapours bearing fluorine and other gases brought about an alteration of the feldspars and led to the formation of other minerals, notably tourmaline. Simple weathering alone would not form china-clay rock, but under the action of weather the granite would disintegrate to yield the sand and clay which contribute to the formation of sedimentary rocks.

The Carn Menellis granite mass, with the neighbouring smaller areas of Carn Marth and Carn Brea between Falmouth and St Ives, is still lower than that of St Austell. Showing similar scenic features, its chief interest lies in the working of tin which has probably been carried on in the neighbourhood since the Bronze Age. Ores have also been exploited in the other granite areas, tin being the most abundant material worked, though ores of copper, lead and zinc also occur. The tin occurs in the form of cassiterite (oxide of tin), which is found in veins and lodes extending in many cases from the granite into the surrounding killas or slates. The veins in any area are usually parallel to one another, and represent fissures which have been filled by deposits carried by vapours emanating from the granite just before its final consolidation. The vapours consisted largely of steam at high temperatures together with fluorine, boron and other gases. In some parts of most of the Cornish granite areas, tin has also been obtained from alluvial deposits in the flats of the rivers and streams. This ore, known as stream tin, represents material produced by the weathering of the veins. Many of these alluvial tracts have been dug over by tin-streamers leaving heaps of debris and water-filled hollows; such deposits were amongst the first to be worked. There are now few rich deposits of stream tin remaining unworked, and most of the mining for many years has been in the veins which run in the solid rock.

The most westerly of the Cornish granite masses is that of St Just, which forms the coast for miles east and north of Land's End, so that the greater part of its margin is washed by the sea. Its narrow eastern edge is marked by a wide depression connecting St Ives Bay with Mount's Bay. The highest land in the mass is just over 800 feet, but much of the land has been planed off at just above 400 feet and is part of the extensive coastal plateau of Cornwall to be described shortly. The cliffs are mainly of granite, although some killas and greenstone are found. The granite cliffs are at their best near and to the south of Land's End, where they exhibit columnar jointing (due to the conditions under which the rock cooled) and have a castellated appearance.

Thirty miles to the south-west lie the Scilly Isles, about 140 in number, of which five are inhabited. These are of granite, low and flat-topped. There is much sand from erosion of the granite and surrounding rocks, and a till has been recognized, showing that once an ice sheet stood here.

Lundy Island must also be mentioned. This lies about eleven miles off the Devon coast at Hartland Point, and is made almost entirely of granite, carved into a plateau form with steep cliffs. This granite, however, is much newer than those discussed above; it is of Tertiary age.

South Devon and Cornwall

In considering the major downfold of Devon and Cornwall, we have described the Devonian rocks of the north, and younger Carboniferous Culm Measures in the centre. The southern part of the peninsula is again made up of mainly Devonian rocks, but they are different in character from those of the north. While, like the northern rocks, they contain many slates and some sandstones, the middle part of the sequence is often made up of massive limestones, and volcanic rocks also occur. Some rocks older and younger than the Devonian are also found, and the granite masses just described are intruded into the area or abut on to it. With this added diversity of rock types still more varied scenery has been developed, especially at the coast. The greater variety of Devonian rocks in the south than in north Devon may be due to their being formed in deeper water farther from the coast than those of the north.

It is more convenient to start this description from the west, for here the scenery is mostly cut in slates, varying in colour but often dark or weathering to rusty brown. Inland, the rocks have been planed off by a number of nearly level surfaces. Particularly well marked is the coastal plateau at about 430 feet above present sea-level, already noticed near Land's End in the granite. This is almost certainly a wave-cut platform and may accord with features at a similar height in neighbouring areas, such as South Wales, and the hills reaching 400 feet in the London basin. Another well-marked surface occurs at about 200 feet: this is particularly noticeable around Torquay, and is almost certainly similarly wave-cut, after the formation of the 430-foot surface.

These inland plateaux end for the most part in cliffs, often in the slates, locally known as killas. The cliffs extend for many miles along both north and south coasts, from Boscastle and Newquay to near St Ives and from Looe to Falmouth. The irregular structures in the contorted rocks give rise to much variety in the cliff patterns, but the close-set cleavage planes produce jagged edges on projecting points and stacks. Vegetation grows rapidly in every small crevice, and though the cliffs are dark and rough, they have a richness of colour which is enhanced by the deep blue of the sea (Plate 35).

Plate 35. *Cornish cliffs, in north Cornwall between Tintagel and Boscastle. The indented coastline caused by the attack of the sea on varying rock types and structures is illustrated, as also is the plateau-like nature of the country inland. These plateau-like surfaces have been planed across rocks folded like those in Plate 33. The cliffs here are about 200 feet high.*

These killas cliffs are interrupted at the western tip of Cornwall. The granite mass of Land's End just described forms the western-most cliffs, and in the shelter of this mass are found St Ives Bay on the north and Mount's Bay on the south, connected by the lowland drained by the river Hayle. St Ives is one of the most picturesque of the fishing towns on the Cornish coast, with delight-fully jumbled groups of cottages built of massive granite blocks. For this is a stone area, and there are very few buildings which are not built of granite or other igneous rock or of the baked killas. Granite is quarried at many places and has been used for gate-posts and beams as well as for ordinary building.

One of the best-known features of Mount's Bay is St Michael's Mount, the steep rocky islet some half-mile from the shore of Marazion, with which it is connected by a causeway. It is a small granite area, a Land's End in miniature, from which the sea has removed the surrounding killas on all sides but the north. Probably it has been finally detached from the mainland only within the last few hundred years, and it may formerly have been, as its Cornish name implies, 'the hoar rock in the wood'.

In the shelter of the granite cliffs both St Ives Bay and Mount's Bay have areas of blown sand. The dunes on the east of St Ives Bay are the more extensive, and though they are now fixed by the growth of vegetation, they have formerly overwhelmed much good land. Penhale Sands, north of Perranporth, is another extensive area of fine sand-dunes on the north coast of Cornwall.

West of Mount's Bay, near Porthleven, is the rather unusual shingle bank of Loe Bar, much of the shingle being of flints derived from Chalk. This bar blocks the Loe Pool, a small drowned estuary into which the Helston river drains. The water then finds its way into the sea by seeping through the shingle bank.

The Lizard and Falmouth

Projecting farther south than Land's End, the Lizard peninsula differs in many respects from all the rest of Cornwall. Its cliffs have perhaps more beauty of colouring than those of any other part of England; the lovely, rocky bays of Kynance Cove, Mullion and Coverack, backed by rugged cliffs of the richest greens and purples, are exeedingly attractive. Briefly it may be pointed out that these dark cliffs are of igneous origin and represent a complex mass of great age, possibly dating from the Pre-Cambrian. The main part of the Lizard rock is of unusual composition: it differs from granite in having no quartz and little feldspar, for its minerals are more 'basic' in composition and darker in colour. Even these, however, have been greatly altered subsequently, and the rock now consists largely of the mineral serpentine. Used extensively as an ornamental stone, this warm-toned rock was formerly sent far afield for architectural purposes. Locally it has been widely used in buildings, for it has the advantages that it is extremely durable and can easily be trimmed. The villages thus acquire a rather sombre green tint, but some of the churches are built of light-grey granite. Most of the area lies between 200 and 300 feet above sea-level, and it is clear that even these ancient rocks have been planed across to form part of the coastal plateau. Inland, therefore, the scenery has less attraction, the soil is thin and often poor, supporting a characteristic flora which includes the Cornish heath (*Erica vagans*).

In the shelter of the Lizard is Falmouth Bay, cut back like Mount's Bay into slaty rocks. But the rivers entering Falmouth Bay are more considerable than those farther west and their drowned valleys form long estuaries. The winding branches of the river Fal extend far inland. Truro has acquired a nodal importance from its position where routes cross the estuary, and above it the river flows on its alluvial plain. Nearer the sea, the wide, deep inlets shelter many attractive villages, such as St Mawes, St Just and Mylor. These long estuaries are the result of another stage in the story of changing sea-level in this part of England. During glacial times, much more recently than the planing of the coastal plateau, the sea was 100 feet or so below its present level long enough for the rivers to carve deep, narrow valleys to that depth. With the rise of the sea to its present level these deep estuaries were drowned, giving the long winding inlets we see today. The channels have been partly filled with sediment, but are still deep enough to provide port facilities suited to different ships, as at Falmouth, Plymouth and Dartmouth.

The resistance of the rocks in Devon and Cornwall has allowed these estuaries and the coastal plateaux into which they are cut to survive, like similar scenery for example around Milford Haven. In the areas of softer rocks, for example around the Mersey and Dee, any coastal plateau there may have been has been eroded away and the estuaries largely infilled, partly by glacial drift, so that the scenery is very different.

The Coast from Plymouth to Dartmouth

The cliffs from Falmouth to Looe are cut in killas and have already been described. East of Looe is Whitesand Bay, with fine beaches and cliffs of highly disturbed pink, purple and green slates. In the Plymouth district we see the Devonian limestone peculiar to south Devon, most readily in the cliffs of Plymouth Hoe, although most of the cliffs both east and west of the Sound are made up of slates and grits, varying greatly in colour and structure.

The rivers here have cut deeply into the coastal plateau: for some way from the coast there is little land above 400 feet. These river valleys and the long estuaries of the drowned river mouths contribute much to the beauty of the Plymouth district. The rivers reach the sea by way of narrow, picturesque inlets; they are not perhaps so remarkable for their beauty as that of the Dart, but winding gracefully between green banks, the Tamar, St Germans River and the Cattewater have many attractive old villages built above their wharves. The character of the Plymouth estuaries can be seen from Brunel's Saltash railway bridge or the Tamar suspension bridge. The Yealm valley has higher banks and, though shorter, has much of the loneliness of the Dart as it flows from Dartmoor.

The coast between Bolt Tail and Start Point is cut into complex and highly resistant very old rocks, chiefly schists of various kinds. These rocks break into rough crags and stand out in irregular rugged points and cliffs. Some of these schists are green (as in part of Bolt Tail), but different shades from almost a pale yellow to a dull green, altered in places by weathering to red and brown, give a peculiar range of colour to these cliffs.

However tough these rocks may be, they have been planed across at about 430 feet as in so many other areas. Through the middle of this mass runs the drowned Salcombe–Kingsbridge estuary, into which no large river drains at present, although the river Avon flows south from Dartmoor in a deep meandering valley to within about two miles of Kingsbridge, before turning south-west to its own comparatively insignificant estuary.

The ruggedness of the cliffs here and elsewhere is smoothed in places by straight slopes of 'head', a mixture of clay and rock fragments, which were shattered by frost in times of periglacial climate, and sludged downslope under freeze–thaw conditions. The base of the cliffs is also notched by raised beach platforms, cut at a time when the sea was slightly above the present level. These are particularly well seen around Start Point. The old village of Hallsands in Start Bay was built on a raised beach platform, protected by a shingle beach. The removal of this shingle by dredging, to provide concrete aggregate for an extension of Devonport dockyard, left the village unprotected, and it was destroyed in a storm in 1917. The assumption of the experts, with the exception of the local geologist, Hansford Worth, was that further shingle would come ashore to replace that removed by dredging. In this case the experts were proved wrong.

The Hallsands shingle beach was one of a series which still survives in Start Bay, the largest being the misnamed Slapton Sands. This is composed mainly of flint shingle, and fronts the silting lagoons of Slapton Ley. These beaches probably originated as barriers in much the same way as Chesil Beach (p. 122), but on a much smaller scale and facing east up-channel instead of the open ocean.

Along the coast east and west of the mouth of the Dart, between Sharkham Point and Blackpool Beach, are steep cliffs formed by grits and hard slates, reinforced in many places by igneous rocks. These produce wild, dark cliffs much like those of Cornwall, while the Dart estuary itself is a deep, narrow inlet in the coast.

The Dart Valley

The valley of the Dart shows several other features of south Devon scenery, particularly in its lower parts, after it leaves Buckfastleigh and Totnes to open out gradually before winding under steep wooded crags to the sea at Dartmouth. Like many rivers of the south coast of Devon and Cornwall, the Dart has been drowned by the comparatively recent submergence of the coast, which has converted its lower reaches into a long estuary. Evidence of this lowering of the land may be found in the submerged forests of Tor Bay and many other places, and in the fact that the Dart estuary is now more than a hundred feet deep, the bed marking the channel having been carved during the low sea-level. Before this submergence the valley must therefore have been even more steep-sided than at present, a deep trench in the low plateau. The influence of such long, narrow inlets on communication between east and west is obvious; the lowest bridge towns like Totnes and Truro, generally some miles above the mouths, have become of great importance, while the coastal tracts between the valleys have been more isolated than would otherwise have been the case. Many relatively un-spoiled pieces of country along the coast owe much to the drowned valley mouths.

The Dart valley gains much in beauty from the fact that it cuts across rocks of several types. Between Totnes and Dittisham it flows for seven miles through a series of volcanic lavas and other igneous rocks formed during Devonian times. These are responsible for much of the striking beauty of the scenery, for each igneous rock stands out in a bold, wooded crag, while the softer beds give rise to more gentle slopes or are marked by the entry of small tributaries. Opposite Dittisham the river widens out still more where it reaches a belt of soft slates, but although it shortly passes off the volcanic rocks, it flows between steep banks in which much grit is present. Such tributaries as Old Mill Creek have etched out the softer beds. The lower Dart estuary below Dittisham is cut deeply between high rock walls, while the river ignores a lower and easier way to the sea in Tor Bay below Ashprington.

The Coast Scenery of Torquay

Many of the characteristic features of the irregular coast at Torquay, the amazing contrast of colour, the dainty bays and bold headlines, result from the complex arrangements of the rocks. The cliffs north and south of Oddicombe Beach are of Devonian limestone, which continues by Babbacombe to Anstey's Cove (Fig. 44). Here the massive beds of limestone, grey or tinged with pink, dip at steep angles and form cliffs which in some ways resemble those of Mountain Limestone coasts. Oddicombe Bay has been cut back

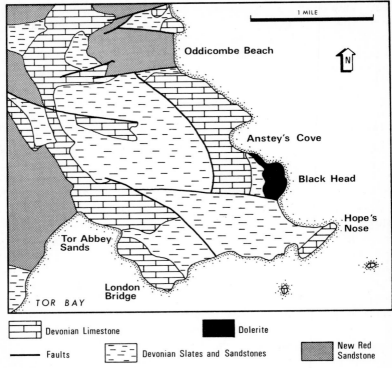

Fig. 44. *A generalized geological map of the Torquay headland.*

between projecting limestone masses along a belt of New Red
Sandstone, let down by faults which almost coincide with the ends
of the beach. These red beds are nearly horizontal and introduce a
different type of cliff scenery, reminiscent of the coast near Dawlish
and Teignmouth.

South of Anstey's Cove the cliffs take on another form, for in
Black Head dark igneous rocks, somewhat akin to basalt, give rise
to a tough mass which resists erosion more successfully than the
tracts of slate which bound it. Hope's Nose is another headland
formed by the limestones, which west of this again form the grey
cliffs where the waves, working along weaker beds, have cut the

arch known as London Bridge. Along these miles of beautiful coast which encircle Torquay, the limestone thus makes up all but one of the projecting headlands, whereas slates and sandstones have been carved into small bays. The same is true also of Tor Bay itself, for this repeats on a large scale the essential features of Oddi-combe. Its northern arm is the complex promontory of Torquay, its southern a great tract of limestone extending through Brixham to Berry Head, and much of the bay is backed by New Red Sand-stone. Low cliffs of these horizontal sandstones between Torquay and Paignton have also been worn through to form a natural arch, but in this case the erosion is along a joint plane.

Above the coast the most striking feature in the Torquay area is the even-topped plateau formed by the limestones both at Babba-combe and, more extensively, above Brixham. With a height of approximately 200 feet, this plateau surface cuts acrosss the various rocks and structures, just as do the coastal plateaux elsewhere in the region. In a similar way, less regular hills rise out of it like islands from the sea (Black Head, Kilmorie Hill and Warberry Hill at Torquay). Deep gorges such as Combe Gorge, south-east of Hele, have been carved into the plateau by rivers. The valleys have re-mained as gorges because of the comparatively short time (in geological terms) since the platform was formed, and the resistance of the rocks.

The Devonian limestone may be likened to the Carboniferous Limestone, for example of Gower, in its coast scenery. It shows further similarity in its solubility, leading to the formation in it of caves. Several of these have yielded evidence of their prehistoric occupants, but Kent's Cavern, Torquay, is of outstanding interest in this connection. The limestone is the predominant building stone in Torquay, as it is also in Plymouth; like the Mountain Limestone it gives rather grey effects, but it is somewhat warmed by delicate tints of pink.

Near by, in an area around Bovey Tracey, is a basin of much newer, Tertiary, rocks, comparable in age with some of those of the Hampshire basin. This unexpected feature is a result of the Alpine folding of Europe, the results of which were expressed as far away as this by faulting, letting down newer rocks into a fault-bounded basin in the older area. Much of the present surface of the basin is flat and heathy, and the deposits are worked in places for clay, A much smaller patch of Tertiary rocks is found at Petrock-stow, on the north flank of Dartmoor.

22. The Chronology of the Changing Scene

Many of the features in English scenery which have been described are of great antiquity. Geological history covers exceedingly long periods of time, however, and it is important to recognize that there are great differences in the dates of origin of these ancient features. Some reference has been made to the relative ages of the different rocks, of plateaux and of river valleys. In this chapter an account is given of the sequence of the chief events in the geological history of Britain, in order to make possible a better appreciation of these degrees of antiquity.

In this account little reference is made to periods of years, although the age of the earth has necessarily interested geologists a great deal. It would be impossible, however, to find agreement in the placing of the sequence of events alongside a table of dates, even an approximate table showing only tens of millions of years, and there is no need to attempt it at present. Geological time is more conveniently divided into periods, the order of which is well established; the history of some of them is known in great detail. Our present position regarding the chronology may be likened to a knowledge of history in which all the important periods such as Norman and Tudor are well known, but in which no dates are fixed, though the approximate duration of each can be estimated fairly closely.

The approximate minimum age of the earth is well established from evidence obtained from the study of the radioactive minerals in rocks of the earth's crust; it is of the order of five thousand million years.

The history of the earth can be divided into two parts, and it may be said at once that we know comparatively little of the earlier (and

CAINOZOIC	QUATERNARY	HOLOCENE PLEISTOCENE	Post-Glacial. The Ice Age.
	TERTIARY	PLIOCENE MIOCENE OLIGOCENE EOCENE	S.E. England submerged in parts. A widespread uplift. Alpine folding. Sands and clays deposited in south-east.
MESOZOIC	CRETACEOUS	CHALK GREENSANDS GAULT WEALDEN	Marine deposits in a sea covering much of England. Fresh-water deposits in south-east England.
	JURASSIC	OOLITES, etc. LIAS	Marine deposits of limestone and clay.
PALAEOZOIC	TRIAS	KEUPER BUNTER	The New Red Sandstone formed mostly in inland lakes. Some desert sandstones.
	PERMIAN	INCLUDING MAGNESIAN LIMESTONE	A period of mountain-building and rock folding — Hercynian or Armorican.
	CARBON- IFEROUS	COAL MEASURES MILLSTONE GRIT	Deposits mostly in shallow water.
		MOUNTAIN LIMESTONE	Limestones formed in a sea covering much of England.
	DEVONIAN AND OLD RED SANDSTONE		Marine deposits in Devon and Cornwall, red beds in fresh water in Hereford and South Wales. A period of mountain-building and rock folding — Caledonian.
	SILURIAN		The rocks of Wales and the Lake District were formed, mostly as marine deposits. In Ordovician times, much volcanic activity.
	ORDOVICIAN		
	CAMBRIAN		
	This line represents a time at least 600 million years ago		
	PRE-CAMBRIAN OR ARCHAEAN		This period may have lasted for at least 3,000 million years. During this time the old rocks of Anglesey, Shropshire, and the Malverns were formed.

Fig. 45. *A chronological table of the main episodes referred to in the growth of England and Wales.*

longer) portion. The events of the later part of the earth's history are more completely known, however, and that division of time may be subdivided into a sequence of periods which are named in Figure 45.

The chief reason for the greater confidence with which these groupings are made, as compared with the vagueness in the earlier part of geological time, is to be found in the occurrence of fossils. Little reference has been made to fossils in the preceding chapters, but it must be emphasized that practically all determinations of the ages of rocks and their grouping into periods is based on the fossils found in them. Any systematic interpretation of geological history or comparison of the rocks in different areas must have reference to their fossils, for the nature of animal and plant life on the earth has of course changed throughout geological time, some forms becoming extinct and new forms being evolved. It is not the purpose of this book to trace this organic evolution, but it may be pointed out that in the more ancient rocks practically no fossils are found, perhaps partly because no life existed in the earliest times; partly because the first forms of life had no skeletons which could be preserved as fossils. Many of these old rocks, moreover, have been exposed to very great increases of temperature and pressure, and have been much modified since the time of their formation.

Thus while it is known that in Pre-Cambrian times there were periods of igneous activity, with volcanoes ejecting ash and lava, and that there were periods of quiet sedimentation when sandstones and muds were laid down, it is not possible to give a satisfactory account of the sequence of these events. Pre-Cambrian rocks occupy great areas in the Highlands of Scotland but they do not contribute greatly to the present surface of England and Wales : they are found chiefly in Wales (in Anglesey, Caernarvonshire and Pembrokeshire), in the Welsh borders (in Shropshire and the Malverns), and more rarely in the Midlands (notably in Charnwood Forest).

Since the Pre-Cambrian there have been several periods when most of Britain has for long been beneath the sea and has received deposits which record the changes taking place. At other times our area has been upraised and mountain chains have stretched across it; then deposits have been formed only, if at all, in small areas, filling lakes and basins or valleys with material worn from the mountains.

With the Cambrian began the first great marine episode of which detailed knowledge is available. It was introduced by the submergence of a land area which had been carved from the ancient rocks, and in the newly developed sea the Cambrian, Ordovician and Silurian rocks were deposited. Most of these rocks consist of sandstones and shales or slates (originally deposited as muds), but there are some thin limestones, especially in part of the Silurian. The sea in which these rocks were laid down varied somewhat in its extent, but through most of the time it covered the whole of Wales and stretched northwards over southern Scotland. The position of the shore-lines varied, and the rocks in these areas show the greatest diversity in their sequence, the Ordovician rocks being the least widely distributed. During the Ordovician also there occurred a series of volcanic outbursts, especially affecting the areas where North Wales and the Lake District are now situated, when lavas were poured out from numerous submarine vents.

These volcanic episodes, however, were followed by a further period of tranquil deposition on a subsiding sea-floor, and it was not until the end of the Silurian that the essentially marine conditions were brought to an end by a great revolution caused by the upraising and compression of this thick mass of sediments. This mountain-building period probably lasted for a long time, but its chief effect was to produce an entire change in the aspect of Britain. The area where there had been sea became a great range of mountains, the folds of the rocks trending north-east to south-west, a direction which is still the dominant 'grain' of the structures in the Lake District and in North and Central Wales, and over large parts of Scotland (whence comes the term Caledonian, for this period of mountain-building).

In Devonian times the only area where marine sediments were laid down lay to the south of the these mountains, occupying what is now Devon and Cornwall. Here marine muds, sands and limestones were laid down to form the rocks now seen in north and south Devon. Over the rest of Britain the newly formed mountains were subjected to rapid denudation and the only deposits were formed in more or less isolated inland basins on the flanks of the mountain ranges, the largest area being centred on Herefordshire. These rocks are mostly red in colour, and constitute the Old Red Sandstone; they are of the same age as the marine Devonian but are of quite different appearance.

With the beginning of the Carboniferous there was once more an invasion by the sea, which spread northwards across England and submerged all but a few areas. The Carboniferous or Mountain Limestone was therefore deposited over much of England and a good part of Wales; in some places it was laid down upon the Old Red Sandstone, but as the latter was very limited in its distribution, the limestone in many areas overlapped it and came to rest directly on the older rocks, giving the striking unconformity which was described in the Lake District. (See also Figs. 23 and 27.)

The Carboniferous deposits accumulated to a great thickness and spread over an even wider area as time went on. Increasing amounts of sediment were carried into the sea to form the deltas which make much of the Millstone Grit. Later, the area became one of predominantly fresh water, with shallow lagoons and swamps overgrown from time to time by dense vegetation, and in these conditions the Coal Measures were formed.

Once more the end of a long period of steady accumulation of sediment was brought about by earth movements, and pressure, this time from the south, bent the newly formed rocks into a series of folds trending east to west. The effect of this pressure was most marked in the south, and the complex folds seen in Cornwall and Devon testify to its intensity (Fig. 33); the east–west trend of the structures and of many surface features in the south-west of England, in the Mendips and in South Wales is due to this folding. Farther north the rocks were much less disturbed (Figs. 23 and 27). This is referred to as the Hercynian or Armorican folding.

Once more the earth movements produced an entire change in the aspect of the region, which became chiefly an area of desert-like conditions with small patches of water in inland basins or in nearly isolated arms of the sea. Here again red beds were laid down, forming the New Red Sandstone. As the mountains were worn away the area receiving deposits increased, and the sediments came to lie unconformably on Carboniferous or older rocks, as in the area near Bristol (Fig. 42) and in Charnwood Forest (Fig. 21).

When a wide tract of desert and salt lake had been established and much of the mountain country had been brought down to a common level, the sea invaded the area. Most of England and South Wales then received deposits of mud and limestone, the Jurassic rocks, in a sea of wide extent. At times there were changes in the position of the shore-line, but a much greater change occurred towards the end of the Jurassic when an uplift of western England and Wales restricted deposition to a small area in the south and east; here the Purbeck and Wealden beds were formed in a freshwater lake. At that time the upraised Jurassic rocks, tilted gently to south-east or east, were exposed to denudation and were rapidly removed, so that when the sea once more spread over the whole region, as it probably did when the higher Cretaceous rocks were laid down, these came to rest on Jurassic and other rocks of various ages.

The Chalk represents a deposit which was formed slowly in a sea of clear water. Over a considerable area the conditions seem to have been remarkably uniform, but the deposition of the Chalk was brought to an end by the rising of the sea-floor. The uplift was probably greatest in the west, but was sufficient everywhere to convert practically the whole British area into dry land.

It will be useful to consider the probable appearance of England and Wales at that time. It must have been essentially an area of very subdued relief, gently sloping towards the east and south-east. Over the greater part of it there can have been little variety of rock type, for the newly formed Chalk certainly had a greater extent than at present. It has been thought that it completely buried most of Wales, but even if it were not so extensive and if much of Wales had remained above sea-level during the time the Chalk was being deposited, it is most likely that this land had been planed down by weather and rivers almost to sea-level.

In the succeeding periods of the Tertiary some deposits were formed in parts of Britain, but these occupy relatively small areas, and the sea did not again cover much of England and Wales. Thus in the early Tertiary, Eocene and Oligocene rocks consisting mostly of deposits of sand and clay were laid down in a sea or estuary which extended over south-eastern England; these rocks are now found in the London and Hampshire basins.

Elsewhere in England and Wales, however, this period was one of denudation, during which rivers flowed down the dipping surface of the Chalk, in an easterly or south-easterly direction into the early Tertiary seas. It is possible that some fragments of this ancient drainage system survive in the west, although the chalk cover has now disappeared, for example the superimposed rivers of South Wales (Chapter 18) and the Dee at Llangollen (Chapter 19). It is very probable that during this part of early Tertiary times, the land which lay to the west of the Eocene seas was planed down to a surface of low relief by such rivers, and it has been claimed by some authorities that the highest summits in the West Country, Wales and northern England represent uplifted portions of this surface (sometimes referred to as a summit plain). Since it was formed before the major earth-movements and faulting of mid-Tertiary times, however, the early Tertiary surface must perforce be greatly deformed and fragmented.

There is no doubt that the Chalk suffered denudation before the deposition of the succeeding Eocene rocks, for Professors Woold-ridge and Linton showed that the surface on which these rocks were laid down (known as the sub-Eocene surface) is bevelled across various layers of the Chalk. Other Mesozoic rocks too were subject to intense erosion, so much so that they survived only in the downtilted south and east and in downfaulted basins in the north and west. In a few lowland areas these newer cover rocks were eroded sufficiently to expose the older basement rocks, as at Charnwood Forest (Chapter 10), but in the uplands these newer rocks were totally destroyed.

In northern Ireland and western Scotland this was a period of volcanic activity, but although some igneous rocks were intruded as dykes in north-western England and Wales, and the granite of Lundy Island was formed, it is unnecessary to deal with these events in this book.

There are no Miocene deposits in England and Wales. This period marked the climax of the folding of the Alps and this mountain-building had great influence on Britain. In the south-east the folds of the Weald, the Isle of Wight and the London basin were fully developed in this period (Fig. 5), while minor folds or faults can be traced through Salisbury Plain and as far west as the Bovey Tracey basin (Chapter 21). Elsewhere the effects of the Miocene movements are not so easily traced, but probably they were of much importance, and many faults may have been formed (or movements may have taken place along old fractures) at this time, especially in the Bristol Channel area. It is believed that the Lake District finally received the domed form on which its rivers were developed, and that early Tertiary peneplains were uplifted to form the high plateaux, although these must have been still further elevated at later dates.

Indeed the Miocene saw Britain assume essentially its present structures. Most of its river systems then began to develop along the lines they occupy at present, and the Lower Thames drainage was collected into the newly formed trough. Probably many east and south-east flowing rivers were longer at that time, and river capture has since modified many courses, new tributary streams etching out the strike of softer beds and so developing such escarpments as those of the oolite and Chalk.

After the major earth-movements and faulting had ceased, a period of intense denudation in late Tertiary times culminated in the formation of a series of peneplains which have now been uplifted to various elevations. In the west these can best be seen in Central Wales (Chapter 17) where the oldest and highest is known as the High Plateau. The lowest and youngest can be traced extensively throughout southern Britain, where it forms the summits of the Cotswolds and the highest chalklands, and may correspond to the Low Peneplain of Wales. Owing to the lack of major earth-movements and faulting since Miocene times these late Tertiary surfaces are undeformed, in contrast to any earlier surfaces. It was the river valleys of these later Tertiary surfaces which were to be greatly overdeepened by the succeeding glaciers of the Ice Age in such places as the Lake District and Wales.

Most of the scenic features of England and Wales have thus developed since the Miocene, but their pattern was determined by events of much greater antiquity. In the case of the highest summits, however, some of the features may be more ancient, dating perhaps from the early Tertiary (or perhaps earlier), while to some extent late Tertiary denudation has uncovered 'fossil' landscape features which had long been buried under the cover rocks, as at Charnwood and perhaps in Anglesey.

During the Pliocene and early Pleistocene, England was still joined to the Continent, and received deposits along the East Anglian coast on the shores of the gulf which occupied much of the area of the North Sea. At that time most of England and Wales was lower by several hundred feet in relation to sea-level than at present. A series of marine platforms, especially well seen around the Cornish coasts, in South Wales, Lleyn and in Anglesey, marks this period of time. The position of sea-level controlled river erosion also, and many rivers carved wide plains sloping gently to the sea. The highest and presumably oldest of the marine platforms, at some 600 feet above sea-level, has had all of its old beach deposits destroyed by subsequent erosion except in south-east England. Here the deposits of red shelly sands and beach pebbles (known as the Red Crag) have been carefully mapped by Professor S. W. Wooldridge, who showed that they were related to a former sea-level which was some 600 feet higher relative to the present sea-level. In East Anglia, however, the Red Crag deposits lie not at 600 feet but at or below high water mark, which demonstrates the amount of local downtilting that has gone on here since early Pleistocene times, associated with the deepening of the North Sea basin.

The lower platforms, whether cut by rivers or the sea, have subsequently been uplifted, probably in several stages, and now are the most striking features in the landscape of many coastal areas. Following their elevation, rejuvenated rivers have once more been enabled to cut vertically and have incised new and deeper valleys in them. As a result of this elevation, many meanders such as those of the Wye were cut and many other gorge-like valleys formed.

The next phase in geological history, representing some two or three million years and ending only ten thousand years ago, was the Great Ice Age of the Pleistocene. The effects of the different ice sheets which on several occasions covered nearly all Britain north of the Bristol Channel and the Thames valley have already been described. They produced many changes in the scenery, although some of them are quite superficial. The form of valleys was greatly modified, especially in such mountainous country as the Lake District and North Wales; in the lower parts of valleys and on the plains great masses of ice-borne material were left, altering the character of many areas and adding new land around some coasts. The diversion of rivers, the cutting of meltwater channels either beneath or marginal to the ice sheets, and the formation of temporary lakes with their associated overflow channels were other results of the Ice Age. Many of these old channels are now either dry or carry a tiny 'misfit' stream, the present volume of which is obviously incapable of carving these wide and deep valleys. Even in areas beyond the ice sheets, periglacial activity, with its frost-shattering and enormous degree of solifluxion, played an important part in landscape modification. Thus despite the relatively short duration of the Pleistocene, the scenery of England and Wales was considerably changed at this time.

As the ice disappeared and the climate improved, vegetation spread northwards across England, and much of the country was covered by forests. During this period the sea-level rose, as water, formerly trapped in the form of ice sheets, returned to the oceans and drowned valleys near their mouths, forming narrow inlets such as Milford Haven and the mouths of the Fal and Dart. Where valleys were wider the drowning gave rise to wide, shallow bays like the Wash, and the silting-up of such areas has lately made them land again.

Man has probably been living in England since Pliocene times, but it was not until the close of the Pleistocene that he began very greatly to affect the landscape. With the establishment of settlements and the clearing of forests, the spread of cultivation gradually altered the aspect of the country. But man's activities in the last 150 years have produced vastly greater effects and man may now be regarded as an important agent of geological change. It is at least a gain that his appreciation of the beauty of his surroundings is increasing while much that is lovely remains.

Glossary of Some Technical Terms

Note: A large number of the technical terms used refer to the 'periods' of geological time, e.g. Jurassic, Cretaceous. These are best understood by reading Chapter 22 and looking at Fig. 45.

Agglomerate: A rock of volcanic origin, composed of irregular blocks of various sizes, comprising solidified lava and fragments of the rocks through which the volcano has broken. Met with mostly in the pipes or necks of volcanoes.

Anticline: An arch or upfold in the rocks, generally produced by the bending upwards of the beds under lateral pressure. Anticlines are structurally weak, and the upper part rapidly becomes worn away (see Figs. 24 and 33).

Arete: A sharp mountain ridge, usually produced by the headward recession of two opposing corrie walls.

Barytes: A common mineral; barium sulphate ($BaSO_4$).

Basalt: A dark lava of basic composition, containing the minerals feldspar and augite, and sometimes olivine. Mostly crystalline but of fine grain.

Base-level: The lowest level to which a river system can lower a land-surface, usually the sea-level of the time, but occasionally and temporarily a hard band of resistant rock.

Beds, bedding: Sedimentary rocks tend to break along planes parallel to that of their deposition, and thus form the beds or bedding of the rock. See also *Joints.*

Boss: A large mass of igneous rock, intruded into and disrupting other rocks. Generally nearly circular in plan; often of coarse-grained rock such as granite.

Boulder clay: A deposit laid down under an ice sheet. Frequently a mixture of clay with boulders, but may contain very little of either, and therefore preferably called *till.* May be tens of feet thick and spread over hundreds of square miles.

Breccia: A composite rock consisting of angular fragments of older rocks cemented together with various minerals such as lime.

Brockram: A cemented breccia or scree of angular limestone fragments in a sandy matrix.

Caledonian: The mountain-building period which dates from Middle Ordovician to Middle Devonian times.

Cassiterite: Oxide of tin; occurs in square prisms capped by pyramids. Commonly blackish-brown to black.

Chert: A hard, flint-like rock, originally a sand, in which the sand grains have been cemented together by the deposition of silica from solution.

Cirque (Corrie or Cwm): An amphitheatre or armchair-shaped hollow, usually excavated on a mountain-side; the slopes are precipitous, the floor nearly level. Due to glacial erosion. In Britain they are commonest on north-east facing slopes, especially in Wales and the Lake District.

Clay-with-flints: A deposit found covering higher levels of Chalk in southern England. It is a residual deposit formed from solution of Chalk itself (the calcium carbonate being removed by solvent water) and also from a denuded cover of later Eocene rocks. It yields heavy, brown, wet, acid soils, and frequently carries damp woodland.

Clitters: A residue of rock waste, frequently including some large blocks, on a gentle slope after denudation has reduced the original surface. Common on the granite moors of Devon and Cornwall, frequently near tors.

Confluence: Junction of two streams, or of a tributary to a main river.

Conglomerate: a rock which consists of an aggregate of pebbles or boulders in a matrix of finer material: a pudding stone. Formed by rapid streams and powerful currents.

Consequent (or dip) streams: A stream which flows in the direction of the dip of the rocks on which it was initiated; the direction of the stream is a 'consequence' of the original inclination of the surface (see Fig. 3).

Cyclothem: A rhythmic sequence of sedimentation, repeated several times, which illustrates a slow spasmodic subsidence. Often found in the Coal Measures.

Denudation: The combined action of weathering, erosion and transportation, to reduce or dissect the existing land-surface.

Dip: The tilt of a bed of rock along its direction of steepest inclination, measured in degrees from the horizontal; also the direction of this dip.

Dip slope: The long, gentle slope of an escarpment, following in general the dip of the rocks (see Figs. 1 and 3).

Dolerite: A dark-coloured igneous rock, basic in composition and resembling a basalt, but composed of rather larger crystals. Generally occurs in a small intrusion such as a dyke or a sill, where it crystallized from a molten state.

Dolomite: Sometimes known as Magnesian Limestone; composed of the double carbonate of magnesium and calcium: $(Ca,Mg)CO_3$.

Drumlin: A low whale-backed ridge of glacial boulder clay, thought to have been fashioned beneath an ice sheet. The long axis of the ridge is generally parallel with the direction of former ice movement.

Erratic: A rock which has been transported by ice a great distance from its source, and has been left stranded when the ice melted.

Escarpment: A unit of scarp and dip slope. *Cuesta* is an alternative name.

Fault: A dislocation in the rocks, where one side has moved relatively to the other. Many fault planes are nearly vertical, others inclined at small angles. Faults have been produced by a variety of causes: some are due to intense lateral pressure, others to differential uplift or to tension (see Figs. 11 and 27).

Fault-scarp: A scarp arising along the line of a fault, caused by the differing resistance of the rocks on either side of the fault, the more resistant rock being eroded more slowly, and hence standing up.

Feldspar: A mineral group comprising various complex silicates of alumina and potassium, sodium and/or calcium. Abundant constituents of most igneous rocks.

Ferruginous: Containing iron in some form; usually as an oxide; descriptive of rock.

Flagstone: A rock that splits readily into slabs suitable for flagging (as in pavement construction).

Flint: A siliceous rock, grey or black; composed of very minutely crystalline silica. Occurs mostly as nodules in the Chalk.

Freestone: A quarrying term to describe a stone easily quarried and which can be dressed or cut in any direction into blocks of any size.

Gabbro: A coarsely crystalline rock of plutonic origin (see *Plutonic*). It differs from granite in so far as its constituent minerals are more basic than acid.

Gneiss: A metamorphic rock, crystalline and coarse-grained, somewhat resembling a granite but showing a more or less banded arrangement of its constituents.

Granite: An igneous rock, composed of crystals of quartz, feldspar and mica, of coarse grain. Results from the slow cooling of a large molten mass.

Granophyre: An igneous rock of medium-sized grain in which the quartz and feldspar crystals are intergrown.

Grit: A rock much like a sandstone, composed mainly of grains of quartz. The grains are either more angular or larger than those in sandstone.

Gypsum: A mineral, commonly formed under arid climatic conditions with strong evaporation. Also known as alabaster ($CaSO_4 2H_2O$).

Head: A mixture of clayey material and frost-shattered stones, formed in an earlier periglacial climate, and which under these conditions moved down quite gentle slopes, thereby causing some rounding off of slopes and cliffs.

Hematite: One of the principal iron ores: Fe_2O_3.

Hercynian: The mountain-building period which dates from Permian times.

Igneous rock: A rock formed by the consolidation of molten material. The characters of the rock depend mainly on the composition of this material, and on the conditions under which it cooled. Material ejected as lava generally cools quickly, material intruded in small masses less quickly, and material in large masses very slowly: the quickly cooled material may be glassy or very finely crystalline (e.g. basalt); the slowly cooled material is coarsely crystalline (e.g. granite). Thus most igneous rocks are crystalline.

Incised meander: Part of an old meander which has become deepened by river incision due to uplift of the land or a fall of sea-level.

Inlier: A mass of older stratified rocks showing through the surrounding newer strata.

Ironstone: A sedimentary rock characterized by its ferrous carbonate content. Common in the Lower Jurassic.

Joints: Divisional planes which traverse rocks perpendicular to the plane of their deposition, cutting them in different but regular directions and allowing their separation into blocks. They are due to movements which have affected the rocks, and to shrinkage or contraction on consolidation. See also *Beds, bedding.*

Kame: A steep-sided ridge or conical hill of stratified sands and gravels formed by meltwaters marginal to a down-wasting ice sheet.

Karst: Irregular limestone topography characterized by underground drainage; a streamless, fretted surface of bare rock which is honey-combed with tunnels; created by solution of the limestone by ground water.

Kettle-hole: A depression in glacial drift, often enclosed, made by the melting out of a formerly buried mass of ice and the subsequent collapse of the overlying drift.

Killas: A Cornish term for the intensely crushed and folded, slate-like sedimentary rock which occurs in that area.

Limestone: A sedimentary rock consisting mainly or almost entirely of calcium or magnesium carbonate, derived from the shells and fragments of former organisms, and deposited originally in water.

Magma: A molten fluid generated at great depth below the surface and thought to be the source of igneous rocks.

Marl: A calcareous clay. *Marlstone.*

Meltwater channel: A deep gorge or valley, now frequently streamless, which was carved out by glacial meltwaters below or marginal to an ice sheet. (See *Overflow channel.*)

Metamorphic rock: A rock which may originally have been either igneous or sedimentary but which has undergone such changes since the time of its formation that its character has been considerably altered: in extreme cases it may be difficult to ascertain its original nature. Heat and pressure are the chief agencies of metamorphism. The commonest rocks of this class are gneiss and schist. *Metamorphosed.*

Mica: A mineral group comprising complex silicates of iron, magnesium, alumina and alkalis. Several different types can be recognized. Mica can be split into exceedingly thin, flexible plates. An important constituent of granite and other igneous rocks, and of some sedimentary rocks.

Mineral: A constituent of a rock, either an element or a compound of definite chemical composition. Mineral as used here has not the same meaning as the 'Mineral Kingdom', which principally includes rocks. Rocks are aggregates of one or more minerals and usually occur in large masses or extend over wide areas.

Moraine: An accumulation of gravel and blocky material deposited at the margins of an ice body.

Nunatak: An area of high land or an isolated peak which protrudes above an ice sheet; it often exhibits widespread periglacial phenomena.

Oolite, oolitic: A limestone of marine origin, composed of more or less spherical grains, each with concentric layers of calcium carbonate, usually formed in shallow water; probably the grains are of chemical origin.

Outcrop: The area where a particular rock appears on the surface. The rock may continue beyond its outcrop, concealed at depth beneath the surface.

Outlier: A mass of newer stratified rocks detached from their main outcrop by subsequent denudation and separated from it by an area of older rocks (see Fig. 1).

Overflow channel (or *Spillway*): A deep gorge or valley carved out by the overflow of a pro-glacial lake at the margin of an ice sheet. These channels are generally streamless.

Peneplain: A land-surface of low relief, worn down by prolonged weathering and river erosion. An extensive peneplain needs for its formation a considerable time during which sea-level remains practically stationary.

Periglacial: A term used to describe those of the cold climate processes and landforms which result from frost action. This term covers frost-shattering, frost-heaving and solifluxion. It is usually most severe at ice-sheet margins.

Platform: An erosion surface produced by marine agencies.

Plutonic: A term referring to rocks of igneous origin formed at great depth by consolidation from magma; the rocks are generally coarse-grained.

Pro-glacial lake: A lake impounded by ice sheets against an area of higher unglaciated ground. The lake disappears either by overflow through the lowest gap in the watershed or by melting of the ice sheet.

Quartz: One of the commonest minerals in the earth's crust. Composed of silica (oxide of silicon). A constituent of granite and sandstones. Crystals are hexagonal prisms terminated by pyramids.

Quartzite: A highly siliceous sandstone. A rock composed mainly of quartz. Breaks with rather a smooth fracture. Usually very tough.

Ragstone: A quarrying term for stone which is difficult to dress, used mostly for rough walling. The Kentish Rag is a calcareous sandstone much used for building in the Wealden area.

Rejuvenated stream: A stream which has received added power to cut vertically into its bed, usually owing to the uplift of the land.

Residual: An isolated hill which stands above the surrounding country because it has resisted erosion to a greater degree. It is the remnant of an older surface which has been almost totally destroyed by the action of streams or marine agencies. Such a hill rising from a sub-aerial peneplain is known as a monad-nock.

Ria: A long, narrow inlet of the sea, caused by subsidence of the land or rise in the sea-level. It is generally regarded as a drowned river valley which deepens gradually seawards.

Rift valley: A valley which has been formed by the sinking of land between two roughly parallel normal faults.

Rock: The geologist does not necessarily confine this term to hard, resistant rocks: he may refer to anything from poorly consolidated sands, gravels and clays to massive, hard stone beds as 'rock'.

Sandstone: A sedimentary rock composed mainly of sand grains, chiefly grains of quartz, cemented together by some material such as calcium carbonate.

Sarsens: Irregular blocks of hard sandstone found on many parts of the Chalk, especially around Marlborough. They are believed to represent remnants of a former indurated sandstone bed, probably of Eocene age, which covered the area. Some irregular masses were tougher than others owing to a subsequent deposit of silica binding the grains together.

Sill: An intrusive sheet of igneous rock, more or less following the bedding of the rocks it invades, but occasionally changing its position among the beds (e.g. the Whin Sill).

Silt: A very fine-grained sediment with particles between $\frac{1}{16}$ and $\frac{1}{256}$ mm. in diameter.

Slate: An argillaceous or clay rock which breaks along planes (cleavage) produced by pressure. The cleavage planes coincide with the original bedding and may obliterate it almost completely. Metamorphic.

Solifluxion: The process of slow flowage of the surface rock waste down-slope when saturated with ground water, frequently under a peri-glacial climate.

Spit: A point of land, generally composed of sand or shingle, which projects from the shore of a water body, frequently at a river mouth or bay entrance.

Stack: A mass or pinnacle of rock left isolated by retreat of coastal cliffs.

Stratum, plural *strata:* A layer or layers of rock, usually referring to bedded sedimentary rocks.

Strike: The direction of a horizontal line on a dipping stratum; level-course of the miner. The strike is at right angles to the dip. The strike direction is generally the trend of the outcrop of the stratum (see Fig. 3).

Subsequent (or *Strike*) *stream:* A stream tributary to a consequent stream or originally so, following the strike direction. Generally along the outcrop of a less resistant bed (see Fig. 3).

Superimposed drainage: A drainage system that has been established on underlying rocks, independently of their structure, from a cover rock which may have disappeared entirely.

Swallow hole (Swallet): A funnel-shaped depression in the surface of a pervious rock (usually limestone) linking with a subterranean stream passage developed by solution.

Syncline: A trough-like fold in the rocks, generally resulting from lateral pressure. In some cases (e.g. the London basin, Fig. 5) the surface reflects the structure, but owing to the wearing of the rocks on either side, many synclines ultimately stand out as hills or mountains (e.g. Fig. 33).

Till: See *Boulder clay.*

Tombolo: A sand or shingle bar connecting an island with the mainland.

Tor: A castle-like rock pile which crowns summits of certain well-jointed rock landscapes, or occasionally occurs at the crest of valley slopes; frequently found in granite but can occur in other well-jointed rocks; its origin is uncertain, with deep-surface weathering, normal sub-aerial weathering or periglacial processes all being suggested by various writers.

Transgression: An advance of the sea over a former land area, due to a change of relative land and sea-level.

Tufa: Material deposited by calcareous springs; usually white or yellowish. Sometimes soft, sometimes hard enough to form a building stone.

Tuff: A rock composed of the fine-grain material ('ash') ejected by volcanoes. Often arranged in beds or layers which have accumulated under water.

Unconformity: The relation of rocks in which a sedimentary rock or group of rocks rests on a worn surface of other rocks (sedimentary, igneous or metamorphic). Frequently the plane of unconformity separates rocks of vastly different ages and the rocks beneath the unconformity are of more complex structure than those above (see Figs. 23 and 42). *Unconformably, unconformable.*

Watershed: The divide between the drainage basins of two rivers (in English usage). American usage defines watershed as the whole of a river catchment basin.

Weathering: The general result of atmospheric action on the exposed parts of rocks. Rain, frost, wind and changes of temperatures tend generally to help in the disintegration of the surface rocks.

On Maps and Books

The reader interested in any special region will no doubt have made himself familiar with the Ordnance Survey maps of the area. He may also feel the need of geological maps. The most useful general map published by the Institute of Geological Sciences (formerly the Geological Survey) is in colour at a scale of about ten miles to the inch, covering Britain in two sheets. Most of England and all of Wales are covered by the southern sheet, Sheet Two, while Sheet One covers northern England and Scotland. In more detail, twenty-three sheets at a scale of a quarter-inch to a mile cover England and Wales, and most areas are also covered on the New Series one-inch maps. These may most readily be obtained by first inquiring of local agents for the sale of Ordnance Survey maps.

The Institute of Geological Sciences also publishes handbooks and memoirs. Regional Handbooks deal with the areas of Britain, and individual memoirs describe in more detail the areas shown on the one-inch maps. Many of these are very suitable for readers wishing to follow up the study of any region, for in most cases they give references to other literature.

Local museums and geological and natural history societies will often be able to supply further advice, while the Geologists' Association (Hon. Secretary, c/o The Geological Society of London, Burlington House, Piccadilly, London W1V 0JU) issues pamphlets giving clear and simple accounts of certain areas of England and Wales, and welcomes interested amateurs as members.

The Geographical Association (343 Fulwood Road, Sheffield S10 3BP) publishes pamphlets for the areas covered by some one-inch Ordnance Survey maps, in the 'British landscapes through maps' series. These give geographical as well as geological information, and include some of the more popular tourist areas.

Local public libraries will contain general geology books. Some of the works mentioned overleaf may be of interest to readers of this book.

H. H. READ and JANET WATSON, *Beginning Geology*, Macmillan and Allen & Unwin.

MICHAEL SMITH, *Essentials of Geology*, Philip.

J. GILLULY, A. C. WATERS and A. O. WOODFORD, *Principles of Geology*, Freeman.

J. F. KIRKALDY, *General Principles of Geology*, Hutchinson.

A. HOLMES, *Elements of Physical Geology*, Nelson.

J. A. STEERS, *The English Coast and the Coast of Wales*, Fontana.

L. D. STAMP, *Britain's Structure and Scenery*, Fontana.

A series of local geological guides is published by David & Charles, Newton Abbot, Devon.

Index